ISBN 978-1-5284-1310-7
PIBN 10908763

1 MONTH OF
FREE
READING

at

www.ForgottenBooks.com

By purchasing this book you are eligible for one month membership to ForgottenBooks.com, giving you unlimited access to our entire collection of over 1,000,000 titles via our web site and mobile apps.

To claim your free month visit:
www.forgottenbooks.com/free908763

English
Français
Deutsche
Italiano
Español
Português

www.forgottenbooks.com

Mythology Photography **Fiction**
Fishing Christianity **Art** Cooking
Essays Buddhism Freemasonry
Medicine **Biology** Music **Ancient
Egypt** Evolution Carpentry Physics
Dance Geology **Mathematics** Fitness
Shakespeare **Folklore** Yoga Marketing
Confidence Immortality Biographies
Poetry **Psychology** Witchcraft
Electronics Chemistry History **Law**
Accounting **Philosophy** Anthropology
Alchemy Drama Quantum Mechanics
Atheism Sexual Health **Ancient History**
Entrepreneurship Languages Sport
Paleontology Needlework Islam
Metaphysics Investment Archaeology
Parenting Statistics Criminology
Motivational

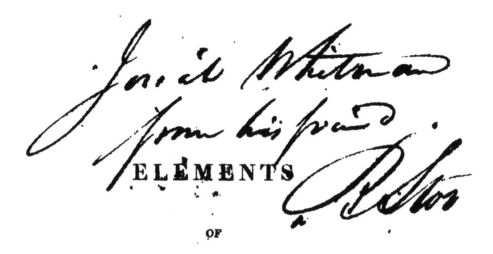

ELEMENTS

OF

CONCHOLOGY,

ACCORDING TO

THE LINNÆAN SYSTEM,

ILLUSTRATED BY TWENTY-EIGHT PLATES,

DRAWN FROM NATURE,

BY THE

REV. E. J. BURROW, A.M. F.R.S. F.L.S.

MEM. GEOL. SOC.

NEW EDITION.

LONDON :

CONTENTS.

It will scarcely be denied, that benefit arises to society from the laudable *fashion*, may I be permitted to call it, of esteeming an acquaintance with the admirable stores of Nature, almost an essential part of polite education. Few would choose to be utterly deficient in information with regard to the higher properties and systematic arrangements of a science, the rudiments of which are now imparted to children with their alphabet; and constitute the basis of the earliest instruction which is bestowed on the youthful mind, to promote its expansion, and direct its first efforts to objects worthy of its choice. The mind of man cannot but be ameliorated by the acquisition of wisdom in all its forms; and if we admit, as surely we must, that there is not any knowledge more worthy of attainment than that of the order and beauty of the Almighty's works, except a knowledge of the Great Creator, and of the relative duties to the performance of which we, his creatures, are bound by so many ties of gratitude and filial love:—if we admit this, it follows that we do not waste our time, when we devote the portion of it which is allotted to relaxation from more severe and necessary employment, to such edifying contemplations, such engaging studies, as those suggested by the book of Nature.

The method which has been adopted, to create a general taste for pursuits, once considered as almost confined to the professional student or the practical collector, has been the very best which could be devised, and has met with proportionate success.

By beginning *ab initio,* by lowering the first steps to the temple of knowledge, by removing the obstructions which first oppose the desire for progress and improvement, implanted for the wisest purposes in every active and ingenuous mind, many persons may be tempted to proceed, till, by a little perseverance, they become masters of the science they otherwise would not have ventured to encounter. For such reasons, the good sense of the well-informed has induced them to write elementary books on the subject of natural history, adapted to the use of those who are less capable than themselves of searching for, and collecting together, all the information which works in different languages afford ; — by this means have been rendered attainable by all the hidden treasures of a most interesting science, the valuable fruits of the laborious investigation of a few.

In each class of the animal kingdom, at least in those which are most generally studied, there is some work, in the English language, explanatory of the terms and fundamental principles ; and in the vegetable and mineral departments, very many excellent helps are offered to the young and inexperienced student : but there is not one* of this description, which is calculated to throw any light upon the difficulties of

* Since the first edition of this work was published, two other Introductions to the Science of Conchology have appeared ; both of which are mentioned in the List of Authors.

Conchology. That some difficulties do exist in this branch of study, will readily be allowed; but at the same time we cannot but regret, that the very acknowledgment of their existence has extremely increased their number: for a supposition seems to have been universally indulged, that conchology lay open as a common field for speculation, in which every individual, whether qualified or not, was at liberty to range, and exercise, without restraint, his genius for invention. The consequence has been, that scarcely two writers on the subject have agreed in their opinions, and that this general want of concurrence has aggravated the evils which each endeavoured to remove.

It is easier to refute error than to establish truth: thus, the several writers who have dissented from the Linnæan school, have, indeed, satisfactorily pointed out some flaws in the great fabric of the " Systema Naturæ; " but in attempting to eradicate the faulty parts, and to supply their place more fitly, they have injured some of the main supports, and have nearly involved the whole edifice in ruin.

The following pages are devoted to the task of facilitating the study of conchology, on the method of the Swedish naturalist; and they are written under the firm persuasion, that a material change is dangerous, even in speculative matters, when the *principle* has stood the test of general consent, and when the means of reaching perfection are not yet, and perhaps may never be, attainable.

Among those who are willing to yield submission to the judgment of Linnæus, whose experience in the scrutiny of Nature's laws is, at least, not inferior to that of any other man, — among those there are probably some, who have been prevented from availing themselves of his method of arrangement, by the want of an elementary introduction, a key to the understanding of his *Catalogue* of Shells, as translated by Dr. Turton, and subsequently corrected and elucidated by Mr. Dillwyn.

It is with a view, in some degree, to supply this deficiency, and to engage a more discriminating attention to the beautiful and interesting collections of shells which are so frequently found in the cabinets of those who know not how to value them, as well as in the superb museums of the metropolis, that the following little treatise has been put together. The scientific reader will find in it very little of novelty, and not any of a theme, which fills too many pages of some conchologists, the invention of new systems, and the demolition of old ones; the disregard of former, and the abuse of contemporary writers. It aspires, indeed, to the approbation of the well-instructed, as an useful manual, and as a correct epitome of the science of Testaceology in its present state. To more than this it does not pretend.

To the great naturalist Linnæus, whose comprehensive mind seems, in many instances, to have anticipated the objections which envy or ignorance would raise against

his system and his fame, we are indebted,
according to the judgment of a large majority, for the most perspicuous arrangement
of testaceous animals which has hitherto
been offered to the public. His system has
been followed in the present treatise, not
because it is considered faultless, or incapable of great improvement, but because,
upon the whole, it is *more* simple and intelligible, and less likely to deter beginners, than any other with which we are
acquainted. At the same time it would be
blamable servility to adhere to the very
letter of the Linnæan law ;—some of the
terms are decidedly objectionable, and are
therefore totally discarded; some few subdivisions of genera are added, which appear
requisite in order to preclude the necessity

of forming new genera, as suggested by certain modern conchologists.

Considerable alterations have likewise been made in the descriptions at the head of each genus, which, being founded on a minute examination and comparison of species, it is hoped may render the technical definitions more clear and satisfactory, and less difficult to be impressed upon the memory.

In the list of descriptive terms, the Latin word has been prefixed to the English translation or derivative; because it may occur to ladies, or to others who follow, as an amusement, the pursuit of collecting and arranging shells, and who may be unaquainted with the dead languages, to wish for a reference to some Latin author.

In such a case they will be enabled, by
means of this small dictionary, to make out
at least the principal characteristics of any
specimen they may be inclined to examine.
For the same reasons, a table of colours
has been considered worthy of insertion,
and has, therefore, been attempted, though
without any very sanguine hope of success.
So much depends upon the modification of
ideas with respect to colours, in themselves
so various, and so few people possess precisely the same impressions of mixed and
blending tints, that it becomes nearly
impossible to establish any thing like a
standard, by which the internal tracings
of a single subject of natural history shall
be accurately and intelligibly described.

The Latin terms of colour are in general

more appropriate than the English; and it would perhaps be better sometimes to Anglicise the word than to translate it.

This is indeed the case with all technical language: it is extremely difficult to retain the precision of Linnæan nomenclature, in rendering it into any other language. It were, therefore, much to be wished, that, instead of endeavouring to erase from their translations all expressions which do not obtain in common conversation, naturalists should rather, by general agreement, adopt a style of phraseology as nearly allied as possible to the Latin,—concise, and appropriated solely to the use of systematic writers. This would tend to diminish, and not to increase, the difficulties which attend the engaging study of natural history in all its branches; for it is surely more easy to remember a well-defined term, though it be new to the ear, than to discover the exact meaning of one which is capable of a diversity of explanations.

Of the Plates it need only be said, that they are all, without any exception, drawn from Nature, and as accurately as an uninstructed hand could with much care accomplish. They are intended to illustrate the *forms* of every natural subdivision of each genus, and will admit of being coloured by the student in conchology, from other specimens, either as an embellishment, or as a very beneficial exercise towards acquiring a more intimate knowledge of this peculiar science.

A few Drawings have been added, with

a view to exemplify the different stages of growth in shells; to point out some among many species of doubtful genus; and to offer to the curious conchologist such rare or non-descript shells as have recently come under the observation of collectors.

ELEMENTS

OF

CONCHOLOGY.

INTRODUCTION.

CONCHOLOGY or Testaceology is the science which treats of the structure, appearance, properties, and methodical arrangement of Shells, the external testaceous covering of certain molluscous animals.

TESTACEA constitute the IIId Order of *Vermes*, the VIth Class of the Linnæan System of Nature.

This order includes all shells of a calcareous nature, that is, composed of carbonate of lime mixed with some gelatinous matter; and excludes those which are called crustaceous, in the covering of which phosphate of lime is a constituent part.

B

Testaceous animals *inhabit* their shells, to which they are only partially attached; whereas the Crustacea are *indued* with theirs, each limb being invested with its own peculiar shield.

Shells are either terrestrial, or found in rivers, lakes, in shallows of the sea, or in the deeper beds of the ocean; they are, however, all subject to the same arrangement, according to their generic and specific characters, and not according to their individual locality.

On the physiology of testaceous *animals*, which belong to a distinct Order, that of Mollusca in the Class Vermes, it will not be expected that much should be said in a mere elementary treatise. Still, only half the object would be attained, if it were totally silent on a subject so fraught with wonder, so indicative of the Wisdom and the Providence which first produced these little admirable architects, and instructed them to form their beauteous receptacles. Much as we delight in viewing the delicate and brilliant tones of colour, the symmetrical structure, or the picturesque and rugged surface which distinguish the different families of oceanic shells, still we experience a more exalted sensation, an inexpressible feeling of surprise and wonder, when we reflect upon the apparently inadequate and helpless agents by which these regular conformations and highly penciled domiciles are wrought. We perceive, at once, that one part, perhaps the chief object, of their destiny is to teach mankind how far superior the humblest worm, when directed in its instinct by infinite and eternal Wisdom, can rise above the boasted ingenuity of that proud and self-sufficient creature, who is permitted to have dominion over the works of nature.

The names will be given, in the Catalogue of Authors, of those who have particularly devoted themselves to the study of the animals, and to their anatomical construction. Some have endeavoured to found a system of Conchology upon the inhabitant rather than upon the shell. This plan has indeed generally been acknowledged as theoretically just, but as uniformly discovered to be defective in the execution, on account of the utter impossibility of procuring, from the unfathomable recesses in which many, if not the majority, vations, depressions, striæ, tubercles, spines, &c. which distinguish individuals in an almost infinite variety of contour, are to be attributed to correspondent projections, tentacula, and other irregularities in the fleshy form of the constructing agent.

It is very important to the inexperienced collector to remark, that the young shells of many species present a very different appearance from that which distinguishes them when in a state of maturity. This difference has caused much confusion among naturalists, who were otherwise well-informed; and, in consequence, the same shells have sometimes been entered in their catalogues under several separate denominations. Not only is the painting altered in the new coats which are laid over the whole surface of the former ones in the progressive stages of their growth, but so *completely* are they changed as not to bear the slightest resemblance, except to a very scientific eye, either in the colours or in the distribution of them. The form varies so essentially as often to deceive the most expert Conchologist, and it is indeed scarcely to be doubted, but that many, which are now universally accepted as distinct species, will hereafter be discovered to be only varieties caused by locality or age. The young shell is often apparently cast in the mould of a genus very different from its own; still there are, in general, some internal or external marks, some obscure characters, which, if previously known and attended to, will prevent the frequency of error. With a view to offer some, though very imperfect, assistance in this respect, a few hints are subjoined of the known variations occasioned by marginal increment, or by superficial depositions of testaceous matter. There are doubtless many more which might be mentioned, and which experience will point out to the careful observer better than any attempt at explanation.

One difference between young and old shells is nearly universal, which is, that the former are thinner, lighter, more transparent, and generally paler in their colours. In figure the multivalves and bivalves vary little, if at all: except that in some genera the mode of increment is plainly discern-

ible by new external layers, extending each beyond
the margin of its predecessor; and in others the
valves appear to be formed of a gradation of suc-
cessive sizes laid one under the other, united only
at the beak, and capable of being separated; constituting however, in reality, as solid and inseparable
a shell as any in which this construction is not visible. Among the univalves, those which are the most
deceptive are the Cypræa, Buccinum, Strombus,
and Murex genera. The young of the others are
chiefly distinguishable by the unfinished edge of
the outer lip : this is either thin, notched, or incom-
plete in its dimensions; still there is no difference
so great as materially to mislead. Some of the
Cyprææ in their young state exhibit the columella
much plaited, the aperture broad, the outer lip expanded, and the spire considerably raised above
the body; they now resemble species of Voluta. In
the progressive stages the columella becomes less
plaited till it be quite smooth, the aperture narrower, and the spire smaller; they then assume
the character of a Bulla, but still they are thin; at
length they gain thickness ; first one lip and then

the other is more or less deeply toothed, and at maturity both perhaps are angular and flat. The spire is then depressed, or rather retuso-umbilicate; and the shells, having undergone as great a variety of changes in the painting as in the form, are at length recognised legitimate Cyprææ. Many Strombi have at first a great similarity to the genus Conus; the winged or lobed lip is wanting; the massy spines are merely tubercles; the body of the shell, instead of being beset with them on all sides, is but slightly undulated, and the sutures are papillary or crenate. Sometimes, though the body and the spire be quite as large as in an adult specimen, there is not the slightest appearance of a tendency to lobes in the outer lip, and even the canal has the direction rather of a Buccinum or Murex, than of a Strombus: in such a case it is extremely difficult, without previous information, to detect the specific character. The Murices have their spines and foliations formed regularly as the whorls increase, and it does not seem probable that they receive any further alteration, after their first construction. The spines, &c. usually increase in size in proportion to the diameter

of the aperture, and are always placed with their
suture or concavity *towards* the mouth.   Thus the
shell is not defective, in general form, at any stage
of growth, each part being, at once, made proportionate and entire : there is, however, a limit at
which the animal, most probably, and certainly the
shell, ceases to be capable of increase, and it is then
only·that the specimen is to be deemed quite com-
plete. The perfecting of the aperture is not effected
in the same way by all of the genus, and therefore
does not admit of an unexceptionable explanation,
but it is very evident to any one who has ever
handled a perfect shell.   If, however, the margin
be more turned outwards than the common direction of the whorls, if it be internally coloured like
the upper surface, finely striate or denticulate, the
probability is that the shell has arrived at its
perfect state.

The following are the Worms of the Order Mollusca, which construct and inhabit calcareous
shells, the objects of our present consideration.

Doris, or (according to Poli) Lophyrus; *Chiton.*
Triton; *Lepas.*
Ascidia; *Pholas, Mya, Solen, Mytilus.*
Tethys; *Tellina, Cardium, Mactra, Donax,*
       Venus, Spondylus, Chama, Arca, Ostrea.
Limax; *Pinna, Conus, Cypræa, Bulla, Voluta,*
       Buccinum, Strombus, Murex, Trochus,
       Turbo, Helix, Nerita, Haliotis, Patella.
Terebella; *Dentalium, Serpula, Teredo.*
Nereis; *Sabella.*

The animals inhabiting the Anomia and Nautilus are either *sui generis,* or have hitherto been
unsatisfactorily described.

It has been generally supposed that the Molluscous animal, of the genus Sepia, mentioned by several naturalists and delineated by some of them, as
the inhabitant of Argonauta, was also the original
fabricator of the Shell. This fact has, however, been

lately doubted; and because the Sepia has shewn
the power of leaving, though not of returning, at
pleasure to its receptacle, it has been considered a
mere parasite. Till the right owner shall have been
discovered, or it shall have been proved that the
tenant uniformly found in possession *could* not be
the builder, we may be pardoned for adhering to
the old prejudice.

———

Before we proceed to the systematic arrangement
and explanation of the Genera, it will be necessary
to acquire some knowledge of Terminology, as
without this their technical descriptions will be unintelligible.

The terms expressive of the forms and characters
of the Univalves are placed first in the following
Catalogue, in conformity with the order adopted by
the author of the " Fundamenta Testaceologiæ "
in the " Amœnitates Academicæ " of Linnæus,
from whence most of them are derived.

THE NOMENCLATURE

OF

CONCHOLOGY.

———◆———

UNIVALVES.

THE SEVERAL PARTS OF UNIVALVE SHELLS, WITH THEIR SPECIFIC MARKS, FORMS, AND DISTINCTIONS.

LATIN.

·ANFRACTUS. The WREATHS or WHORLS; the circumvolutions of the spire around the Columella, are either

ancipites. Two-edged; longitudinally carinate at the sides, or

bifidi. Bifid; divided transversely by a line or furrow, as it were, by a suture.

canaliculati. Channeled; having a small excavated channel along the suture.

carinati. Keeled; the whorls compressed angularly.

contigui. Contiguous; growing together.

coronati. Crowned; surrounded towards the apex with a simple row of protuberances or spines.

distantes. Disjoined; perfectly separate at the sides.

frondosi. Leafy; having the varices spreading into leaf-like or crested forms.

imbricati. Imbricated; covered with scales laid partly one over the other like tiles.

indivisi. Entire; opposed to bifid.

lamellati. Lamellate; plated, surrounded transversely with membranaceous excrescences.

lineati. Lined; engraven with lines, either *raised* or *excavated:* either *longitudinal*, extending from the apex to the base, *transverse,* following the course of the whorls, or *striate*, rendered rough by transverse striæ.

"Lines" sometimes express the tracing only of the colour.

obsoleti. Obsolete; having the suture obliterated.

scrobiculati. Scrobiculate; covered with small pits or excavations.

scripti. Lettered; marked with various characters resembling letters.

sinistri. Left-handed; heterostrophous, turning round the pillar from right to left, instead of pursuing their usual, opposite, course.

spinoso-radiati. Spinosely radiate; beset with spines in a circle, either *concatenate,* united at their bases, or *setaceous,* like bristles.

striati. Striate; encompassed with very fine raised or excavated lines : *punctate* striæ are those which have elevated or impressed points placed along them; the points may be *concatenate,* strung like beads, or *pertuse,* deeply excavated.

sulcati. Sulcated; marked with broader lines, either hollow, ridged, or elevated.

APERTURA. APERTURE, MOUTH; the orifice, entrance, or opening of the shell.

bimarginata. Bimarginate; having the lip with a double margin.

bilabiata. Bilabiate; constructed with both an internal and external lip; in opposition to those shells which are destitute of the interior one.

dehiscens. Gaping; the lower part of the lip being distended.

coarctata. Coarctate; contracted, straight: opposed to effuse.

effusa. Effuse; having the lips separated by a sinus or gutter, so that if the shell were filled with water it would flow out at the back part.

reflexa. Reflex; having the fore part of the lip reflected towards the lowest whorl.

repanda. Spreading; having broad open lips.

resupinata. Resupinate; turned upwards.

transversa. Transverse; situated in a plane parallel to the inclination of the spiral line of the whorls; in the direction of that line.

dentata. Dentate; furnished with teeth.

Apex. Apex, Tip; summit of the spire.

decollatus. Decollate; apparently mutilated, having the spire, or the upper part of it, horizontally cut off.

papillaris. Papillary: opposed to acute, having the apex semi-globular.

ARTICULI. JOINTS; the parts of the whorls, in some of the Nautili, between the genicula.

BASIS. BASE; the opposite extremity to the apex; in some shells that part of the belly which is next the aperture, in others the very lowest point of the beak. In this sense it is either *emarginate,* indented by a deep canal, or *entire,* without indentation.

CANALIS. CANAL; a continuation or prolongation of the aperture along the beak, which forms a gutter by the involution of its sides.

CAUDA, ROSTRUM. BEAK; the elongated bases of the belly, lips, and columella.

abbreviata. Abbreviated; shorter than the lower whorl.

clausa. Closed; having the edges of the canal meeting, or nearly so.

elongata. Elongated; longer than the body.

explanata. Plain: dilated at the margins.

truncata. Truncate; cut off transversely.

COLUMELLA. PILLAR; the middle column about which the wreaths form their spiral circuit.

abrupta. Abrupt: truncate at the base.

caudata. Caudate; elongated so as to project beyond the body.

plana. Flat; extending into a plain lip.

plicata. Plicate; marked with transverse folds.

spiralis. Spiral; caudate and spirally twisted.

COSTÆ. RIBS; keel-like processes reaching from the apex to the periphery of the shell.

fornicatæ. Arched; beset longitudinally with hollow scales.

DIGITI. CLAWS; digitate or finger-like lobes of the outer lip.

DORSUM. BACK; the upper part of the body when laid on the aperture.

EPIDERMIS. SKIN: a membranaceous covering of the shell, found on some but not on all species.

GENICULA. GENICULATIONS; the contraction of the whorls so as to correspond with the internal divisions or dissepiments.

LABIUM. LIP; the internal or columellar mar-
interius. gin of the aperture.

In the Patellæ this name is given to a testaceous membrane in their concavity called *fornicale* when under the vertex, and *laterale* when it constitutes a chamber in the side.

exterius, LABRUM. The outer margin of the aperture.

anticum. Anterior part; next the spire.

posticum. Posterior; that nearest the rostrum.

coarctatum. Coarctate: drawn back to the base.

digitatum. Digitate; divided into attenuated diverging lobes.

solutum. Disengaged; separated from the whorls by a sinus or gutter.

fissum. Cloven; almost divided in the middle by a linear sinus.

mucronatum. Mucronate; projecting in one single sharp point.

OPERCULUM. LID; a plate or door with which some species close the aperture of their shells: it is either of a horny, testaceous, or membranaceous substance, and varies much in shape and contexture.

RADII. RAYS; elevated striæ tending from the centre to the periphery.

ROSTRUM. BEAK, formed by both lips being produced towards the base into an attenuated or narrow process.

SIPHO seu SIPHUNCULUS. SIPHON; a small cylindrical canal perforating the partition in a chambered shell, is either *centralis*, through the centre; *lateralis*, through the margins of the partitions; or *obliquus*, cutting the axis of the whorls.

SPIRA. SPIRE; the upper whorls collectively.

cariosa. Carious; corroded, or, as it were, worm-eaten.

capitata. Capitate; terminated by an obtuse head.

exquisita. Exserted; much attenuated.

plana. Flat; having the upper whorls of an equal height, so that the spire appears truncate.

retusa. Retuse; having the lower whorls of the spire pressed into the body.

retuso-umbilicata. Retuso-umbilicate, the spire being so much impressed as to seem rather concave than convex.

SUTURÆ. SUTURES of the whorls; the spiral line of connexion between them.

duplicata. Duplicate; with a double elevated line or stria.

marginata. Marginate; raised, with a prominent keel.

TESTA. SHELL.

antica. Anterior (part); that which is towards the spire.

clavata. Clavate; club-shaped, thicker above and elongated towards the base.

convoluta. Convolute; when the exterior whorls spirally involve the interior.

corticata. Corticate; covered with an epidermis.

cylindrico-umbilicata. Cylindrically umbilicate; the umbilicus of which is a cylindric cavity.

emarginata. Emarginate; having the margin excavated by a sinus.

exumbilicata. Imperforate; destitute of a hollow umbilicus.

fusiformis. Fusiform; intermediate between conical and ovate, or tapering and a little ventricose.

imbricata. Imbricate; with plaits parallel to the margin.

interrupta. Interrupted; continued by new accretions, or additions of the shell.

involuta. Involute; having the margin of the exterior lip turned inward.

lineis crispata. Wrinkled; rendered rough by flexuous lines.

marginata. Marginate; the sides of the shell being thickened.

obovata. Obovate; nearly oval, but narrower and produced at the base instead of the apex.

perfoliata. Perfoliate; having a horizontal suture girt with a deflex or overhanging margin, as if one shell were placed upon another.

polythalamia. Chambered; divided internally by various partitions.

radicata. Radicate; affixed by the base to some foreign substance.

rostrata. Rostrated; beaked.

spiralis. Spiral; so twisted that an imaginary plane passing through the middle of the outer whorl shall bisect all the others, or divide them into two equal parts.

turrita. Turreted; the whorls gradually attenuating or decreasing in the form of a cone; the length of these shells always considerably exceeds their width.

turbinata. Turbinate; the belly swelling or tumid; the spire comparatively small, rising as it were from the hollow of the body.

umbilicata. Umbilicate; possessing an umbilicus.

VARICES. VARICES; longitudinal gibbous sutures formed in the growth of the shell at certain proportionable distances on the whorls.

continuati. Continuous; running through all the whorls.

decussati. Decussate; longitudinally and transversely divided or crossed.

scrobiculati. Scrobiculate; engraven with cavities towards the margins.

VENTER seu CORPUS. BELLY or BODY; the lowest whorl, the last circumvolution of the shell, which is the most tumid.

VERTEX. VERTEX; in the Patella signifies the supereminent point of the shell.

submarginalis. Submarginal; placed near the posterior margin.

UMBILICUS. UMBILICUS; a hole in the base of the columella visible underneath.

perforatus. Perforate; having a hollow duct or opening through the centre to the very apex.

subobtectus seu *Rima umbilicalis.* Umbilical cleft; where the lip is reflected over the hollow, and only the margin of the opening is apparent.

Margo columnaris. Columnar Margin; when the margin of the columella forms the interior wall of the aperture.

BIVALVES.

Ambitus. Circumference of the shell.

Area. Slope.

antica. Anterior slope or Area; that side of the beaks or that space in which the ligament is situated.

distincta. Distinct; separated from the sides of the shell either by a channel or a keel.

inflexa. Inflex: having incurved lips.

literata. Literate; ornamented with characters like letters.

postica seu Areola. Posterior slope; the other side of the beaks.

marginata. Marginate; surrounded by an elevated margin.

patula. Patulous; the margins of the areola gaping,

serrata. Serrate; saw-like in the channel of the areola.

Apices. Beaks; the tips or extreme parts of the umbones or bosses.

auriformes. Auriform; ear-shaped, having an incurvated arch between the beaks.

corniformes. Corniform; horn-shaped, long, produced, straight, mucronate, or pointed.

inflexi. Inflex; incurved, bending towards each other.

reflexi. Reflex; turned towards the areola.

spirales. Spiral; twisted spirally.

AURICULÆ. EARS; angular processes either on one or on both sides of the beaks.

BASIS. BASE; that part of the margin which is immediately opposite to the beaks or summit.

CAVITAS. CAVITY; interior superficies of the shell.

CALLUS. CALLUS; composed, as it were, of two abbreviated ribs, united at the base, converging from the apex towards the posterior margin.

CARDO. HINGE; that part of the circumference in which the valves cohere, or are attached to each other. It forms the thickest region of the shell, and is inwardly furnished with various protuberances and hollows.

depressus. Depressed; having a simple tooth extending the hinge towards the anterior part.

excisus. Cleft; gaping with a transverse slit.

longitudinalis. Longitudinal; running nearly the whole length of the shell.

lateralis. Lateral; projecting on either side.

reflexus. Reflex; bent or folded over the exterior margin.

terminalis. Terminal; placed at the extremity of the shell.

truncatus. Truncate; having the beaks transversely cut off, and the hinge placed within them.

Costa cardinis. Hinge rib; an elevated line running from the hinge internally to the inferior margin.

CICATRIX. CICATRIX; internal muscular impression.

DENS. TOOTH; an acute projection within the hinge, by which the valves are united and often articulated.

anticus. Anterior; the tooth nearest the ligament, or on the side of the area.

posticus. Posterior; next the areola.

complicatus. Complicated; membranaceous forming an acute angle.

duplicatus. Duplicate; deeply cleft, approaching
to bifid.

depressus. Depressed; turned inward.

erectus. Erect; perpendicular to the plane of the
hinge.

longitudinalis. Longitudinal; produced along the
margin.

masticans. Articulate; furnished with many teeth,
so situated that when the valves are closed
they lock into each other.

primarius seu *cardinalis.* Primary or Cardinal;
situated between the beaks.

Discus. Disk; the middle of each valve, the
convex part between the bosses and the
margin.

Fossula. Sinus; a small hollow in the hinge.
Foveola and *Scrobiculus* have nearly the
same signification.

Intestinum. Intestinal ligature; a membranaceous tube by which some of the

species of the Genus Anomia adhere to foreign substances.

LABIA. LIPS. -

interna. Internal; the small lips within the ligament, which are united by it.

hiantia. Gaping: when the lips are wide, or distant from each other.

retracta. Retracted; opposed to prominent.

truncata. Truncate; shorter than the suture.

externa. External; the margins of the area, or anterior slope about the ligament.

incumbentia. Incumbent; when one of the external lips hangs over the other.

LIGAMENTUM. LIGAMENT; a cartilage which closes the suture, and is affixed between the internal and external lips of the shell, connecting the valves.

LIMBUS. LIMBUS; circumference within the margin.

LONGITUDO. LENGTH, measuring from the summit to the opposite margin; and the breadth, *Latitudo,* in a direction at right angles to this.

LUNULA. LUNULE; a small crescent-shaped depression on either the area or areola.

MARGO. MARGIN; the outline or edge of the shell.

RADII. RAYS; See above.

echinati. Spinous; longitudinally armed with spines.

vesiculares. Vesicular; beset with knots, or prominences internally concave.

RIMA. SUTURE; the interstice which separates the valves when the ligament is wanting, or the hollow which is covered by the ligament.

clausa. Closed; the internal lips being thickened so as to cover the whole suture.

SQUAMULÆ. SCALES.

canaliculatæ. Channelled; excavated longitudinally.

fornicatæ. Arched; concavo-convex.

imbricatæ. Imbricated; laid partially over one another, like the tiles on a house.

tubulosæ. Tubulous; having the sides folded together so as to form a tube.

Striæ. See above.

abbreviatæ. Abbreviated; not extending to the margin.

bifariæ. Bifarious; diverging.

recurvatæ. Recurved; elevated, membranaceous; tending towards the apex.

inæquilineatæ. Inæquilinear; not parallel.

Sulci. See above; sometimes they are used, though perhaps improperly, to signify the same as *Costæ,* Ribs.

fornicati. Arched; covered with arched scales.

Testa. See above.

antiquata. Antiquated; longitudinally sulcate or furrowed, but interrupted by transverse accretions, as if lesser valves were periodically added to the apex or beak.

aurita. Eared; the hinge produced on either side the beaks into a flat prominent angle.

barbata. Bearded; superficially covered with rigid hairs.

compressa. Compressed; the valves flattened, and the bosses less than usually gibbous.

dorsata. Dorsal; the back obtusely keeled.

edentula. Toothless, with reference to the margin.

hians. Gaping; the valves so partially closing, that the margins do not every where touch each other.

inflexa, Inflex; bent towards the anterior slope.

linguiformis. Linguiform; tongue-shaped, linear with the extremities obtusely rounded.

navicularis. Navicular; boat-shaped.

pectinata. Pectinate; longitudinally sulcate or striate; the striæ forming an acute angle in the anterior part.

radiata. Radiate; with diverging rays flowing from the apex longitudinally to the circumference.

rostrata. Rostrated; the anterior extremity elongated and narrowed.

fastigiata. Fastigiate; terminating transversely at the base.

saccata. Saccate; distended towards the beaks.

truncata. Truncate; part of the circumference being abrupt, or, as it were, cut off.

VALVÆ. VALVES.

æquilateræ. Æquilateral; where the anterior and posterior sides of the valve are equal and similar.—Inæquilateral; the reverse.

æquivalves. Æquivalve: where one valve is perfectly similar to the other.

dextræ et sinistræ. Right and left. If the shell be placed upon its base, with the area in front, and the valves be then divided, the right valve will be opposite the left hand of the examiner, and the left valve opposite his right hand.

lacunosæ. Lacunous; marked with a longitudinal depression.

prominentes. Prominent; when one valve protrudes in any part beyond the other.

UMBONES. BOSSES; the swelling parts near the beaks, the highest extreme of which, or that point furthest removed from the base, we consider as the summit.

fornicati. Arched; inwardly much excavated.

MULTIVALVES.

Basis. Base; in Chiton and Lepas, that part of the shell which is affixed to extraneous bodies, either with or without a Peduncle; in Pholas, as in Bivalves, the part of the margin opposite to the summit of the beaks.

Ligamentum. Cartilage; the membrane which connects the valves.

Limbus. Border; the marginal membrane of Chiton.

Operculum. Operculum; in Lepas, the lid which closes the vertical aperture, consisting usually of four triangular valves.

Pedunculus. Peduncle; the tubular support of certain Lepades.

coriaceus. Coriaceous; like leather.

Valvæ. Valves.

dorsales. Dorsal; the solitary valves in the compressed quinquevalve species of Lepas.

succenturiatæ. Accessory; those smaller irregular valves which are annexed to the hinge of the Pholades.

GENERIC SYSTEM.

WE come now to the arrangement of Shells, after the method of Linnæus; and if we bear in mind the professed foundation of that arrangement, we shall find but little cause to complain, either of perplexity in the general scheme, or of want of precision in the several descriptive parts. It is upon *external* characters, upon those of the testaceous covering, and not upon the genus or species of the worm, that we are to erect our system; because the former are the most obvious, and the least liable to misconception.

Premising, then, that by the word *Valve* is meant any single piece of calcareous substance, let the form be what it may, which serves as a habitation, or protection, either partially or entirely, to a Molluscous animal, we shall understand the natural division of all shells into Multivalves, those composed of more than two pieces; Bivalves, of *two* distinct parts; and Univalves, of one entire formation.

Linnæus places them in the above order: and although it may, at first sight, appear more systematical to begin with the Univalves, and proceed to those of more complicated structure; still, upon due investigation, the Linnæan series will, at least, by many, be found most eligible. On this subject there has been, and probably will be, much difference of opinion; yet it is to be considered, that not only the number, but the beauty and importance of the species increase progressively from the Multivalve division, which contains the fewest, to the Univalve, which far exceeds the other two collectively. Much more discrimination and experience are requisite, to point out the varieties, which amount to some hundreds, of Patella, than to arrange the comparatively small genus Chiton. It is not the number of valves, but the number of the different specimens to be examined, which constitutes the difficulty of arranging a genus, and of defining its limits with perspicuous accuracy.

The strong affinities existing between some subdivisions of genera, which, according to prescribed rules, are nevertheless distinct, and the fallibility to which all human systems are too plainly subject, have occasioned, most indisputably, many errors in the classification of the individuals; but this affects only the question of accuracy in a catalogue; and it is not of material importance to the elementary inquirer, whether these doubtful species can, or cannot, be better placed than they are at present. An addition to the number of subdivisions might, it would seem, be sometimes beneficial, and render the work of classification more satisfactory and easy. Those shells which are deficient in some one point alone of their analogy, might, when separated into natural families, or sections of genera, distinguished by those affinities which they do possess, more readily find the proper situation to which they are entitled. But it is pretty evident, that by an increase of genera, you do any thing rather than simplify; and simplicity is surely as much a desideratum in Conchology, as in other branches of natural history. It may be as well to mention here, that the terms Cochleæ and Conchæ, often adopted by writers on this subject, the former to signify univalves, and the latter bivalves, have not been used at all in the following pages, because they appear unnecessary, and are super-seded by more significant and appropriate names.

The generic characters of the multivalves are derived from the situation or number of the valves; of the bivalves, from the hinge; of the univalves, either from the aperture, or conformation of the shell. Generic distinctions must be permanent, that is, invariably existing in all the species which compose the genus.

Specific characters are taken from the colour, surface, or figure, which are different in the various parts, of which each shell, in every genus, is con-structed.

Varieties are formed, usually, by adventitious circumstances, and differ from others of the same species, for the most part, only in the painting, or some non-essential shape.

Multivalves are either *parasitical*, as the genus Lepas; or *unattached* to foreign substances, as Chiton and Pholas.

Bivalves are divided, 1st, into those whose hinge is furnished with internal teeth, either *not inserted* into the opposite valve, as Mya, Solen, Tellina, Donax; or *inserted*, as Cardium, Mactra, Venus, Spondylus, Chama, Arca; and 2dly, those without teeth, as Ostrea, Anomia, Mytilus, Pinna.

Univalves are distinguished into those, 1st, possessing *a regular spire*, which, as to their aperture, are either effuse, as Conus, Cypræa, Bulla, Voluta; canaliculate, as Buccinum, Strombus, Murex; or coarctate, as Argonauta, Nautilus, Trochus, Turbo, Helix, Nerita, Haliotis: 2dly, having no spire, or a very irregular one, as Patella, Dentalium, Serpula, Teredo, Sabella.

The Linnæan genera are thirty-six in number, and are arranged according to the order of the annexed Table, in which are inserted the total amount of species hitherto described in each genus, and the proportion of them which has been found in and around the British isles, according to the last edition of "Pennant's British Zoology."

Many species, doubtless, are known which have not yet made their way into the catalogues of conchological writers, and we have reason to suppose that hundreds or thousands may be still unknown.

No account is here taken of varieties.

MULTIVALVES.

1	Chiton	- 40	8
2	Lepas	- 45	19
3	Pholas.	- 12	5

BIVALVES.

4	Mya	- 41	18
5	Solen	- 35	12
6	Tellina	- 97	25
7	Cardium	- 54	16
8	Mactra	- 37	15
9	Donax	- 21	7
10	Venus	- 116	34
11	Spondylus	4	0
12	Chama	- 25	1
13	Arca	- 45	9
14	Ostrea	- 84	13
15	Anomia	- 32	5
16	Mytilus	- 49	17
17	Pinna	- 21	3

UNIVALVES.

18	Argonauta	11	0
19	Nautilus	- 58	22
20	Conus	- 155	0
21	Cypræa	- 68	2
22	Bulla	- 61	20
23	Voluta	- 186	16
24	Buccinum	172	1
25	Strombus	44	
26	Murex	- 171	2
27	Trochus	- 130	13
28	Turbo	- 167	77
29	Helix	- 189	70
30	Nerita	- 70	11
31	Haliotis	- 21	1
32	Patella	- 102	14
33	Dentalium	15	7
34	Serpula	- 38	28
35	Teredo	- 4	1
36	Sabella	- 25	13

Total - 2445 550

DEFINITIONS AND DESCRIPTIONS

OF

THE GENERA.

———◆———

MULTIVALVES.

CHITON.

(See Plate III. Fig. 1.)

VALVES many; disposed in an imbricated manner, or with the edges one over another, along the back.

The general form of the shell is convex-oval, and the margins of the valves are connected by an elastic, coriaceous cartilage, often scaly, hairy or spinous, which permits the free motion of the valves, or segments, in a longitudinal direction.

The Chiton much resembles the Oniscus *entomon,* or marine woodlouse, and has often been mistaken for it. It is capable, in the same manner, of rolling itself into a perfect ball.

By far the greater number of species consist of 8 valves; and the specimens of 6 or 7 are so rare, that they may, not unreasonably, be suspected of having sometimes been either carelessly or fraudulently *composed* of disconnected valves. The marginal membrane being entire, will always afford an obvious security against such sort of imposition. There are, however, instances of the C. *squamosus* possessing only 7 valves with a perfect margin; but these must be considered as *lusus naturæ*.

The animal adheres, usually, to rocks, or other shells, by means of a gelatinous fluid which exudes from the papillary under-surface of its body. Like the Patellæ, it is generally found parasitical, but possesses the power, of removing from its station.

The name of the Genus, Chiton, is derived from the Greek word χιτών, signifying a coat of mail; and aptly expresses the loricated appearance of the shell, arising from the position of the valves.

LEPAS.

Section A. Sessile. (See Plate III. Fig. 2.)

B. Peduncled. (Fig. 3.)

SHELL multivalve; affixed by the base; valves unequal, erect.

In the first division of this genus, the species have mostly 6 valves, are conical, adhere by their base to rocks, shells, &c. In the second they are compressed, varying in number, supported by a peduncle or hollow membranaceous stalk. The former have also an operculum composed of 4 or 6 triangular valves closing internally the vertex of the shell, and affixed by an elastic cartilage to the interior of the upper margin. The cellular structure of these shells is worthy of observation: the hollow striæ and cavities of the triangular valves are vertical, and the space between them is filled with nearly horizontal filaments, forming a more compact and even substance. The two divisions of this genus, though appearing, at first sight, extremely dissimilar, are yet congeners in their principal essential characters, being multivalve, and parasitical; and the species pass gradually from one conformation to the other in so connected a series, that it would seem needless, and indeed destructive of simplicity and order, to separate or disarrange them. The aperture in the conical division increases proportionably to the growth of the shell, by means of the extension, in width, of the triangular parts between the valves. The peduncled species are called Barnacles, from a strange prejudice formerly entertained, that they were transformed into geese; originating perhaps in their feathery tentacula, and in their being observed moving about on their long flexible tubes above the surface of the water. Both divisions are invariably attached to extraneous substances, usually in groups.

The genus takes its name either from λίπας, the *rock*, to which the shells adhere, or from λεπάς, the denomination bestowed by the ancients on the family of Patellæ, on account, probably, of the *scale*-like manner in which they are adherent.

PHOLAS.

(See Plate III. Fig. 4.)

SHELL bivalve, inæquilateral, divaricate or gaping, beaked; having smaller accessory valves situated upon the hinge and posterior slope. Hinge recurved, furnished with a tooth.

The Pholades, as their name, derived from the Greek φωλέω, imports, *seek a hiding-place* in all descriptions of rocky fragments, and even in wood, piercing the substance while they are young, and gradually increasing the dimensions of their cell according to their growth. The largest species and the finest specimens are most frequently found in chalk, which being the softest of calcareous rocks, admits, perhaps, of a more easy and rapid progress, than the indurated stones in which they are sometimes discovered. It is not, however, yet understood by what instrument they are enabled to penetrate the substance of their future prison; when, judging from the size of the aperture, they are still in a young and, probably, a feeble state.

May not this process be forwarded by chemical as well as mechanic means? May not the peculiar secretion, which, both in the body of the animal, and when separated from it, emits a phosphorescent light, act in some degree as a menstruum on calcareous matter, and woody fibre? If so, the difficulty of procuring an entrance, and of enlarging the internal chamber, would be much diminished.

It appears, indeed, scarcely possible that these Mollusca should be able to obey the instinct of their nature without some aid from a softening or dissolving fluid. They are doubtless deposited, as soon as formed, in the superficial cavities of the rock or wood; for they are usually discovered in great numbers about the same place, as if from the ovary of a common parent.

The accessory valves are affixed to the hinge and upper margin by a gelatinous substance, which is liable to decay after the death of the inhabitant: hence arises the circumstance of their being found very frequently defective.

BIVALVES.

MYA.

(See Plate IV. Fig. 2.)

SHELL mostly gaping at both ends : hinge seldom with more than one tooth, solid, thick, and spreading, not inserted into the opposite valve. (Fig. 1.)

The shells of this genus are, in shape, chiefly broader than long, and generally smooth, or slightly striated. The hinge, in some species, is connected internally by an elastic cartilage, which in the one valve is sunk in a triangular sinus, and in the other is protected by the broad tooth. In others the cartilage is wanting, and the teeth are more or less complicated ; sometimes they are crenate, or lobed, and one or more of the lobes, seldom the principal one, are slightly inserted. This genus is rather to be distinguished by the size and thickness of the teeth, than by any assignable form of them, in which they differ much.

In a few instances the hinge is destitute of teeth,

and the shell is only to be classed by other points of generic resemblance. It would perhaps be better if this genus were recast; for, as it at present stands, it includes, most palpably, several stolen species. The inhabitants of the Mya burrow in the sand or mud, leaving a channel, through which they occasionally thrust their retractile proboscis.

Mya *margaritifera*, a species which is found chiefly in the large rivers of northern latitudes, is known to produce pearls, partial secretions of the same matter which forms the inner coating of the shell, in considerable abundance. The British islands, especially Ireland, have been considered famous for their fisheries of the Mya, and a few pearls of great value have at different periods been obtained from these sources; but the quality of British specimens in general is by no means held in the highest estimation.

The Greek original μύαξ, from μύω, to *compress*, (whence the word Mya is derived,) was formerly applied to the Genus Mytilus.

SOLEN.

(See Plate IV. Fig. 4 and 5.)

SHELL oblong, gaping at both ends. Tooth of the hinge subulate, reflex, often double, not inserted. (Fig. 3.) Lateral margin obsolete.

The valves are convex, and in some of the more oval species, the beaks, which are mostly obsolete, are short and incurved.

In a few species there is an internal rib extending from the hinge to the margin of the cavity in a longitudinal direction. The disproportionable breadth of the shells in this genus is a remarkable character, and many of them are extremely brittle.

The form of the solen is extremely well adapted to the motions of the animal which it contains; these being exclusively in a perpendicular direction. It makes its way through the sand of the sea-shore, to a certain depth, and thence again to the surface to seek for food. This is a tribe which can hardly be confounded with any other.

E

The teeth, though differing in the species, are very unlike those of the Tellinæ, which most resemble some individuals of the genus in external characters.

Specimens of the oval and linear forms are presented in the plate; but they have not been divided into sections, because there exist intermediate species which it might be difficult to exclude from either.

The name Solen is derived from σωλὴν, *a tube*. The Latin appellation Unguis, by which this family has sometimes been designated, was probably given on account of the *nail*-like or horny substance and colour of many of the species. By the French the shells have been called Manches de Couteaux, in allusion to their shape.

TELLINA.

A. Ovate and thickish.   (Plate V. Fig. 2.)
B. Ovate compressed. (Fig. 3.)
C. Suborbicular. (Fig. 4.)

SHELL compressed towards the anterior slope, and often bent. Teeth of the hinge mostly three; lateral teeth plain, or not existing, in one of the valves. (Fig. 1.)

The beaks are very short, and usually lean towards the ligament, which is large, and covers the prominent margin of the suture.

It is difficult to beginners in the science of Conchology to discriminate between the genera Tellina and Venus. In the forms of the shells, one genus much resembles the other; and the teeth of some small species are not distinct to an inexperienced eye. The difference will therefore best be learnt by observing *principally* the inclination of the beaks: if they at all tend towards the ligament, the shell will belong to the genus Tellina; also, if any remote lateral teeth be discernible, if the anterior slope be compressed into an acute wedge-shaped form, or if it be crooked. But the primary teeth are still in either case to be attended to; for they constitute the best, though not always the most obvious, criterion.

It is worthy of remark, that the bend or fold of the extremity of the compressed slope is, without any exception, towards the left valve; that is, towards the right hand of the observer when the shell is placed upon its base with the area in front.

The Tellinæ are found buried in the sand or fine gravel of the sea-shore, some of them in rivers or wet ditches.

Τελέω, from whence Tellina has been derived, places but an ill-defined limit to the tribe, as it will include all shells which *arrive* quickly at *maturity*. It is difficult to imagine why the name was originally adopted, or why it has been retained by almost all Conchologists.

CARDIUM.

A. Compressed. (Plate V. Fig. 7,)
A. Sub-equilateral and turgid. (Fig. 6.)

SHELL sub-equilateral, mostly equivalve, convex, longitudinally ribbed, striated or channelled, margin toothed. Hinge with two primary teeth alternate, both much incurvated, lateral ones remote and inserted in the opposite valve. The beaks are turned inward, and the bosses turgid.—Many of the species in this genus bear the appearance of a heart, when viewed with either of the slopes in front; some very exactly assume, and all more or less approximate to, the cordate form.

One obvious character of the Cardium is, that the ribs are usually if not universally longitudinal, and not concentric or transverse, as they chiefly are in Tellina and Venus, which approach most nearly to the same heart-like outline, and are therefore in their exterior appearance most easily confounded with the Cardium. The disk is usually convex, but sometimes much compressed, keeled, or angular. The ribs and furrows of the two valves are so disposed as to alternate at the margin, and to lock accurately and firmly into each other. Although a variety of shells are as justly entitled to the appellation καρδία, from being *heart*-like, as those which constitute the genus Cardium, still the latter all possess this character, and are distinguished from others by their hinges.

In two species, C. *echinatum* and *aculeatum*, a circumstance is observable, which is altogether anomalous in the structure of either bivalves or univalves; the spines, which are pointed on the anterior and flattened on the posterior side of the valve, have their opening or suture situated towards the apex, and not as in other instances towards the lower margin. There is doubtless some singular construction of the animal which effects this formation of the spines : but for what purpose these shells are made to differ from their congeners, such as C. *Isocardia*, it is not easy to determine.

MACTRA.

(Plate VI. Fig. 2.)

SHELL inequilateral, equivalve. Hinge having the primary tooth complicated, with an adjacent sinus, lateral ones remote and mutually inserted. (Fig. 1.) Shape various, sub-triangular, or broader than long, sometimes gaping. The form of the hinge is very remarkable. The compound middle tooth and the adjoining hollow, filled with an elastic cartilage connecting the two valves, are easily to be distinguished, when their construction has once been attentively observed.

We may here remark, that whenever an internal cartilage does occur, it is generally of a triangular shape, or of one tending to a three-sided prism, the acute angle of which is placed under the apex of the shell. It is not possible to conceive a conformation more beautifully adapted to the motion of the valves and the articulations of the hinge. This cartilage, it appears, should be considered as performing some other functions than those of the outward ligament; for there are few, if any, instances, in which it is found without the latter. In Mactra, in the pectinated Ostrea, and in Mya, there are both; but the internal tendon seems designed to give a firmer connexion and greater tenacity to the valves, where, for want of a sufficiently deep insertion of the teeth, some strong tie is requisite to prevent a lateral motion. Thus it does not exist in Tellina, Cardium, or Venus.

To μάκτρα, *a kneading-trough*, we can trace but little resemblance in our Mactra. The word was, however, no doubt, at first selected for some good reason of which we are now no judges.

DONAX.

(Plate VI. Fig. 4.)

SHELL having the margin often crenulate, and anteriorly very obtuse. Hinge with two primary teeth and one marginal, solitary, remote, rarely either double, triple, or deficient. (Fig. 3.) Ligament external.

The shells of this genus are nearly all triangular, inequilateral, truncate, or concave on the anterior slope, and assume the form of a wedge.

The single primary tooth in one of the valves is bifid. The shell usually gapes a little at each end.

The derivation of the name from δόναξ, *an arrow*, may perhaps have been adopted on account of the rapidity with which the animal darts into the sand, on the approach of danger: his sagittate shell is well calculated for penetrating quickly the yielding substance of the shore.

VENUS.

A. Subcordate. (Plate VII. Fig. 2.)
B. Orbicular. (Fig. 3.)
C. Oval, a little angular on the anterior margin.
    (Fig. 4.)

SHELL having the lips incumbent on the anterior margin. Hinge with three teeth, *all approximate*, the lateral ones diverging from the apex. The area and areola are well defined. In this genus the beaks are uniformly, or with *very* few exceptions, turned towards the posterior slope, that is, *from* the ligament. By this character they are distinguished from many of the Tellinæ, which otherwise resemble them, as well as from some species of different genera which have no *remote* lateral teeth.

The first section A. includes all the shells which are nearly cordiform, as well those which are furnished with spines on the circumference of the area, as those which are merely wrinkled, striated, or smooth. They are very different in external appearance from the species arranged under the other sections, and indeed approach nearer to the form of a Donax. The area is generally large, lanceolate, *distinct*, very often coloured differently from the disk. The areola is smaller, cordate, depressed, and gives a peculiar character to the marginal outline of the valves.

In the orbicular division a heart-shaped lunule is distinctly marked on the areola, and is often prominent; the area is usually compressed.

The spines of V. *Dione* can scarcely be alleged as an exception to the remark made on CARDIUM *echinatum*, that in all other shells the suture of the spines is placed on the side furthest from the apex; for in this muricated Venus the spines are evidently formed by an extension of the transverse plaits at an angle having its base towards the beaks, and therefore necessarily open on the inner side.

To the sea-born Goddess, of the ancient mythology, Linnæus has dedicated the present genus, selected from the produce of her native shores.

SPONDYLUS.

(Plate VII. Fig. 6.)

SHELL inequivalve, rough. Hinge having two recurved teeth with an intermediate hollow; sometimes eared. (Fig. 5.) One of the valves convex and thick, the other flatter.

The intermediate sinus contains a connecting cartilage, as in some other genera. The lower of the two valves of this singular shell is, in many individuals of the first species, *Gædaropus*, produced towards the apex into a projecting beak, one side of which is excavated so as to leave a flat space perpendicular to the hinge. Through this the cartilage is continued in a curvilinear direction to the very tip. The appearance is that of an imperfect specimen, having one of the apices artificially mutilated. The surface is always coarse and rough, tuberculated or spinous, and more or less pectinated. With the exception of the strong and well-defined teeth which are peculiar to this genus, and which are totally wanting in that of Ostrea, the affinity between them is exceedingly great. To the division Pecten, particularly, the resemblance of the longitudinal ribs and eared hinges, is very remarkable. Some species are, however, destitute of ears, and of the exserted beak.

These shells are attached to rocks at considerable depths in the ocean, from which they are separated with the greatest difficulty. They are often bored by Pholades and marine insects.

The animals of the Spondyli are commonly eaten on the coasts of the Mediterranean, in which sea they are found in considerable abundance.

The term σπόνδυλος is sometimes used for the prickly head of an artichoke, though more generally for the vertebræ of the back-bone, to either of which some likeness may be found in the Spondylus.

CHAMA.

(Plate VIII. Fig. 2.)

A. Gaping.   B. Closed.

SHELL thick. Hinge furnished with a gibbous callosity, obliquely inserted in a corresponding channel. This callosity or tooth is either simple or crenate, occasionally double or triple. (Fig. 1.)

Some species of this genus are truncate, and gape on the posterior margin; and though all do not so, they agree in being equivalve, and sub-equilateral, in having recurved beaks, and no prominent lips. The form differs much, being sub-globose, reniform, cordate, or rhombic. No genus, perhaps, contains two species so dissimilar in outward appearance as C. *Cor* and C. *Gigas;* still they are linked together by the Linnæan bond of relationship, the hinge. This genus, indeed, were it formed only upon the name, derived from χήμη, *a gaping*, would exclude a large proportion of its shells, and would properly retain only one species, C. *Gigas,* which is the largest of all testaceous productions, and is sometimes as much as two feet or more in breadth. The oval, or rather lanceolate opening in the posterior slope, is probably made use of by the fish to eject a byssus, with which it fastens to rugged substances and resists the action of the waves.

The very incongruous varieties of external forms, of surface, and of general character, which distinguish the several species of the 2d section, render it impossible to offer a good example in illustration of the family; it has not, therefore, been attempted. Although the margins in this division be not all *perfectly* closed, yet they are *comparatively* so; and are obviously separated merely as a passage for nutriment, not for any substance such as a bunch of adhesive filaments.

ARCA.

A. Margin very entire, beaks recurved. (Plate
    VIII. Fig. 4.)

B. Margin entire, beaks reflected. (Fig. 5.)

C. Margin crenate, beaks recurved. (Fig. 6.)

D. Margin crenate, beaks inflected. (Fig. 7.)

SHELL inequivalve; hinge with numerous sharp
teeth alternately inserted. These teeth are most
frequently placed in a longitudinal direction on
the hinge. Shape various, oblong, rhombic, subcordate, lenticular, subovate, or eared; and the
surface is sometimes covered with a rough cuticle
and beard.

In many species the area lies flatly between the
remote beaks; it is horizontal or a little inclined,
and is covered with the ligament in a manner
peculiar to this genus. In these the line of separation at the hinge is perfectly straight.

The name of the genus is taken from one of its
species, which is supposed to resemble Noah's ark,
Arca Noæ. With this the other species are connected, not by form, but by the construction of
the hinge.

OSTREA.

A. Valves furnished with ears, and radiate;
  Pecten.

  1. Equilateral; ears equal. (Plate IX. Fig. 1.)
  2. Ears unequal. (Fig. 2.)
  3. Valves more gibbous on one side; hinge
     rather oblique; ears short. (Fig. 3.)
B. Rough, and generally plated on the outside.
   (Fig. 4.)
C. Hinge with a perpendicular grooved line.
   (Fig. 5.)

SHELL generally inequivalve, more or less eared.
Hinge without teeth, having a hollow cavity or
sinus, and in many instances lateral transverse
grooves or furrows.

Shape of the pectinated division orbicular, attenuated towards the hinge, eared; of the other
divisions orbicular, lobed, or produced towards the
apex. Much as the divisions A. and B. vary from
each other, there seems more generic affinity
between them than is usually admitted. The
animals may not be specifically alike; but the

F

Linnæan arrangement is professedly founded on the
shells; and therefore, when the hinge is much of
the same description, and the other characters are not
absolutely incongruous, it is surely unnecessary to
divide them otherwise than into sections, according to their natural conformation. Neither, in
fact, are all the varieties of the species in each
division extremely dissimilar; for it is not at all
uncommon to meet with specimens of the common
Oyster, Ostrea *edulis*, of a pectinated and eared
form; so that a regular transition might be shewn
from one to the other family. The third division
C. is connected to the former ones by the O. *Pes
Lutræ*, which partakes largely of the properties of
both; and differs from the 3d subdivision of A.
only in the shape and direction of the groove
which constitutes the hinge, and in the thickness
of the valves, which renders it not unlike the
division B.

Strong locomotive powers have been attributed
to the Pecten, which are, it is said, exerted in a
most singular manner. A very rapid progress is
effected by the sudden opening and closing of the
shell. This is done with so much muscular force,

as to throw it four or five inches at a time. In the water, an equal dexterity is evinced by the animal, in raising himself to the surface, directing his course *ad libitum*, and suddenly, by the shutting of his valves, dropping to the bottom.

His less active relative, the rough Oyster, in the mean while, is contented to remain fixed to his first station, surrounded by an innumerable progeny, continually increasing with wonderful fecundity. His motions consist only in turning from one side to the other, which he accomplishes more by sagacity than any natural agility or inherent strength. He contrives to bolster up one side, by a gradual deposition of soft mud, till he stands nearly upright; then, availing himself of the flowing or ebbing of the tide, he opens his shell, and is tumbled over by the pressure of the water. As expedition is not his object, this mode may answer well.

It has, however, been observed that the young fry possess the power of swimming very swiftly by means of an undulatory motion of the branchiæ.

It is from the words ὄστρεια or ὄστρεα, terms applied generally to Bivalves by the Greek naturalists, that the Latin Ostrea and our *Oyster* are derived.

ANOMIA.

A. Perforated in the apex. (Plate X. Fig. 1.)
B. Perforated in the disk. (Fig. 2.)
C. Imperforate.

SHELL inequivalve; one valve flatter, the other more gibbous, often produced and incurved towards the beak. One or both valves are frequently perforated near the apex. Hinge toothless, consisting in some species of a prominent linear callus or cicatrix, with or without one or two internal marginal teeth, and two bony rays in the cavity.

Shape orbicular or obovate, rarely quadrilateral.

The definition of the Anomia, given by Linnæus, is very unintelligible: it is therefore attempted to render it less so, by altering some parts of the description which appear unnecessary and tend only to obscurity. This genus in fact comprehends species differing so greatly in essential characters, that it is hardly possible to form a description which shall not be contradicted by some among them. The perforation in one or other of the valves, for the emission of a ligature, by which the animal adheres to extraneous substances, is perhaps the most general, though not an universal character.

The name ἀνομία, *transgression of the law*, offers the best apology for the incongruities of the genus. It appears to be the general receptacle of all shells which cannot scientifically be referred to any other family. Of the species which now constitute the genus, 3 divisions have been formed, which may in some degree assist their methodical arrangement, and explain to the learner the principal points in which they differ.

The best example of the 3d section is A. *Sella.*

The animals which inhabit Anomiæ are probably of different species, or even of distinct genera.

The individual whose description Linnæus prefixes to the genus can belong only to one division of it. He says, " *Body* of the animal ligulate, emarginate, ciliate; hairs affixed to the upper valve. *Arms* 2, linear, longer than the body, connivent, projected,—ciliate on both sides, hairs attached to both valves." This specimen is figured in the " *Fundamenta Testaceologiæ,*" Tab. III. Fig. 25.

MYTILUS.

A. Parasitical, affixed by claws. (Plate X. Fig. 3.)
B. Flat or compressed, and slightly eared. (Fig. 4.)
C. Ventricose, or convex. (Fig. 5.)

SHELL rough, often affixed by a thick byssus, or silky beard. Hinge toothless, distinctly marked (except in a few species) with a subulate line, excavated longitudinally. Shape either folded, crested, lobed, or attenuated towards the apex.

The Mytili, though not all absolutely parasitical, or inseparably attached to other substances, are all rendered stationary by some mode or other of adherence: the silky filaments emitted by some species, are entwined in the corallines and stones at the bottom of the sea, and securely anchor the groups of muscles which are found there. Some perforate the rocks, and larger shells, and form to themselves a habitation, from which, like the Pholades, they possess no means, nor perhaps inclination, to escape.

The Pearl-bearing shell of the Indian fisheries is the Mytilus *margaritiferus:* it is most abundant

and in greatest perfection on the coasts of the
Persian Gulf and of the Island of Ceylon. The
term Pearl-*oyster* is, therefore, incorrectly, though
commonly, applied to the shells which principally
produce pearls; for although they may sometimes be found in the Ostrea *edulis* and other species, the Mytilus *margaritiferus* and the Mya *margaritifera* are pre-eminently those from which
the pearls of commerce are obtained. In the great
fisheries established to supply the Eastern market,
the number of fish annually taken up from their
beds by divers, whose perilous trade it is to search
for them, is almost incredible. Some of the shells
contain one or more pearls; others, not any. They
are usually detached, but adhere often to the valves,
and are extricated by opening the shell and wash-
ing. After the day's work, the pearls which have
dropped out are selected and assorted. The
small or seed-pearls are worth from three to seven
guineas per ounce. Those of half a grain weight
are sold for about eighteen pence or two shillings
each; and those of one grain from three to four
shillings; of two grains from seven to nine shillings each; of five grains from thirty-five to forty-

two shillings; those of eight or nine grains, if fine in colour and shape, are of arbitrary value. The finest specimens, of extremely rare ·occurrence, have fetched enormous prices, and have even been considered invaluable, fit only to adorn the regalia of princes and contribute to the costly splendour of Asiatic potentates. These beautiful and unassuming productions, so delicate and varied in their tints, so elegant in their forms, are more highly estimated and more generally used as ornaments in Asia than in Europe, and consequently the most precious are retained by the Asiatic merchants.

The word Mytilus is from μυτίλος, the diminutive of μῦς, *a mouse*, to which little sharp-nosed animal some of the small dark species might have been compared.

PINNA.

(Plate XI. Fig. 1.)

SHELL sub-bivalve, brittle, erect, gaping, throwing out a beard or byssus. Hinge toothless, the valves being inseparably united.

Shape broad at one end, and gradually tapering towards the other. Valves convex, equal, and connected on the side of the hinge by a membrane, in such a manner as to form in fact an univalve shell, bearing the appearance of a bivalve. The valves are incapable of motion in their hinge, but are liable to a forcible separation.

In the one instance of the Pinna, the method of Linnæus in making the hinge, or that part nearest the apex, the base of a bivalve shell, seems unquestionably derived from the habits of the animal, which stands erect under water, infixed in the mud by the smaller end of his habitation. But we may doubt whether, according to the usual definition, that part of the margin to which the ligament adheres ought not to be considered as the hinge: if so, the length of the shell will be less than its breadth; which is contrary to the Linnæan descriptions of the several species. If we then define the base to be the side opposite the hinge, consistently with the plan in which we have ventured to differ from the great Swedish naturalist, the Pinna will be supposed to rest on one end, and not to stand upon its apex. Such a supposition is at least justified by the analogy of the Solen, which is known to perforate the sand laterally.

Πίννα, sometimes written πίνα, the Greek designation of the genus Pinna, may have been originally derived from πίνος, *the dirt*, or *soil*, in which the shells of this family are found.

Ælian amuses us with a story of the Pinna being accompanied by a crab, who lodges and boards with him; and who, by way of return, when a fish chances to swim within his precincts admonishes his friend by a gentle nip. The Pinna then opens his valves and admits the head of the unlucky fish, who, thus entrapped, is converted into supply for their joint larder.

From this genus, so equivocally placed between the two divisions of Testacea, consisting of one, and of two parts, it is but a short remove to the next link in the admirable chain of nature.

UNIVALVES.

ARGONAUTA.

(Plate XII. Fig. 1.)

SHELL univalve, spiral, involute, membranaceous, unilocular or without chambers. Aperture cordate.

This delicate and brittle shell, composed as it were of two parts, of the two sides connected by the keel, approximates more nearly to the bivalves than any other univalve, and is therefore placed foremost in the last division of Testacea.

The Nautilus of Pliny has been separated from the chambered genus, bearing that name in the Linnæan System, and has been denominated Argonauta, from 'Αργοναυτης, a companion of Jason in the celebrated voyage of the ship Argo. The art of navigation is supposed to have owed its origin to the expert management of this instinctive sailor. He was observed by the ancients, and subsequent experience has confirmed the observation, to raise himself to the surface of the sea by ejecting a quantity of water, and thus diminishing the specific gravity of his vessel. When floating in a calm, he would throw out two or more tentacula, to serve as oars. If a favouring breeze sprung up, he would stretch a fine membranaceous sail on two extended limbs, and, steering with his other arms, shew his ready skill in naval tactics, by numberless evolutions with his fragile bark. On the approach of danger he would suddenly haul in his tackle, and, by a rapid absorption of the water, betake himself to the security of the fathomless abyss, his native dwelling. On account of this talent of quickly discerning and avoiding his pursuers, the sagacious little mariner is seldom taken in the act of sailing, but is usually fished up upon rocky shores, or entangled in the nets of fishermen at sea.

One species, the rare and beautiful A. *Vitreus*, has sometimes been placed among Patellæ, but, surely, with little judgment. It has, also, been made to form a new genus, but without necessity, for it answers correctly to the Linnæan definition of Argonauta, and moreover possesses the prominent characteristic, the dentate keel.

NAUTILUS.

A. Spiral, rounded, with contiguous whorls. (Plate XII. Fig. 2.)

B. Spiral, rounded, with separated whorls. (Fig. 3.)

C. Elongated and straightish.

SHELL univalve, convolute, smooth, many-chambered, the divisions perforated and connected by a continued siphunculus or pipe, formed of a thin, testaceous matter, and lined with a membrane of the animal. The dissepiments are convex inwardly, and the chambers become gradually larger from the tip; in the last or external one of which the animal is supposed to fix his habitation, keeping up a communication with his interior apartments by means of the hollow tube which passes through them all.

This and the preceding genus have many natural characteristics common to both; but they are distinctly separated by the circumstance of one being concamerate, the other not so. The inhabitant of the Nautilus, as the shape of the shell will indicate, possesses the power of floating on the surface of the sea, but is more frequently found reversed, and bearing his boat upon his back. The animal of the Nautilus *spirula* possesses a conformation analogous to that of the Cuttle-fish, and in the posterior extremity of his body he bears the shell, which is only partially uncovered. This animal was brought from New Holland by M. Peron. It has been presumed that other concamerated shells are similarly situated, and that they serve merely to protect a small portion of the animal with their exterior chamber.

No exemplification is given of the division C., because all the shells which it contains are either minute or fossil, neither of which come within the design of a work merely elementary, and confined to recent subjects. Fossil conchology, though subject to the same arrangement, is well worthy of a separate consideration.

To ναυτίλος, *a sailor*, this genus evidently owes its appellation.

CONUS.

A. Spire nearly truncate or flat. (Plate XIII.
    Fig. 1.)
B. Pyriform. (Fig. 2.)
C. Elongated. (Fig. 3.)
D. Ventricose, contracted at both ends.
E. Thin, ventricose. (Fig. 4.)

SHELL univalve, convolute and turbinate. Aperture effuse, longitudinal, linear, toothless, entire at the base. Columella smooth. Base attenuated, often marked with oblique rugose striæ. The aperture is sometimes dilated. The whorls are mostly flat, often channelled, rarely crowned. The superior beauty of this genus renders it highly interesting, and the conical form distinguishes it from all others, except the Voluta and Trochus, the former of which has a plaited, and not a smooth columella; the latter a transverse, and not a longitudinal aperture; and the conoidal form is erect, and not inverted.

The shells of this genus are usually covered with

an epidermis, and under it, when in good preservation, bear the most brilliant polish. This fine surface contributes much to heighten the delicate and glowing tints which are diffused over some of the finer species in an infinite variety of undulations, clouds, spots, bands, and reticulated figures.

In enumerating the divisions of this genus, as they stand in Gmelin's edition of Linnæus's Systema, the comparative length of the spire, with that of the body, has been suppressed; because no two species answer exactly to the same measurement, and even in the same species the proportion will be found to differ. The character, therefore, can only create confusion. The division D. contains but one species, C. *Sinensis,* which is not figured, because the author has not been able to procure an inspection of the shell itself; and he is unwilling to deviate from his purpose, of offering none but original drawings.

The Greek κῶνος· expresses the peculiar form to which the genus is indebted for its name.

CYPRÆA.

A. Mucronate, or with a projecting spire. (Plate XIV. Fig. 1.)

B. Obtuse, and without manifest spire. (Fig. 2.)

C. Umbilicate. (Fig. 3.)

D. Margined. (Fig. 4.)

SHELL univalve, involute, subovate, obtuse, smooth. Aperture effuse at both ends, linear, toothed on both sides, longitudinal.

The genus is remarkable for the high polish which adorns it in its native state. The only species of other genera which are likely to be confounded with it, are one or two Bullæ: these, however, have only one lip toothed or slightly plaited. The outer lip is usually thicker, and more incurved in this than in any other genus, resembling more or less the inner one.

A very remarkable and unprecedented property has been ascribed to the Limax inhabiting the Cypræa; namely, the power of quitting his tenement, and elaborating a new one more suited to his ne-

G

cessities. It is supposed that, during the operation
of perfecting his first receptacle, he himself increases in dimensions, till it be with difficulty that
he is contained within it: when this is the case,
he squeezes himself through the narrow aperture,
perhaps with considerable pain, as the whole
cavity of the shell was insufficient for his ease;
and committing his unprotected body to the briny
element, is, doubtless, agreeably irritated, till his
secreting powers are enough awakened to lay the
foundation of a new testaceous covering. These
circumstances are rendered still more extraordinary, by observations which may be made upon
the genus, very much at variance with them. Had
they not, indeed, been announced as from ocular
demonstration, they might be suspected of having
originated in some accidental occurrence, or imaginary habit.

In the first place, analogy concludes against any
such wandering propensity; for it is not even sur-
mised that the congeners of the animal possess it;
and it is difficult to conceive why this Limax alone
should grow disproportionably, after the comple-
tion of his first shell. We find, in general, that

in the construction of the Univalves, the whorls increase in magnitude as they advance transversely from the apex, and receive no increment after the perfect aperture is formed. This bespeaks a limit in the growth of the inhabitant, at which he is taught by instinct to finish off his work. If, therefore, for any reason, upon which from ignorance we cannot argue, these slugs do quit their shells, it is not, surely, on account of an anomalous excess of bulk. A remark which has been made by the author, whether justly or not will hereafter be decided, would seem to favour a very opposite opinion. In numberless specimens of the Cypræa *exanthema* which he has inspected, the thinner and obviously younger shells were, with few exceptions, of large dimensions, whereas the fully ocellated and thicker individuals were considerably smaller. So much difference is there usually in size, that they might be considered as two distinct varieties, were not a progressive course easily obtained, from those having broad bands alone, to those covered with white spots, and totally destitute of bands. This course, however, follows an inverse law with regard to magnitude. The above remark is simply proposed to the attention of naturalists, without attempting to account for the fact adverted to.

The Cyprææ live deeply buried in the sand, from whence it is said, at the full moon and during its increase, these little lunatics, for such they have been already proved, crawl forth to expatiate upon the rocks, and leave there their shells for the benefit and instruction of Conchologists.

To the Cyprian Goddess this genus, including certainly some of the most beautiful of species, is very fitly, but, as to its orthography, very ill inscribed. From the original Κύπρις will come Cypria, but not Cypræa.

BULLA.

A. Birostrate, or produced both ways. (Plate XIV. Fig. 5.)
B. Caudate. (Fig. 6.)
C. Without any elongation. (Fig. 7.)
D. Tapering. (Fig. 8.)

SHELL univalve, convolute, unarmed. Aperture subcoarctate, oblong, longitudinal, entire at the base. Columella oblique, smooth.

The diagnosis of this genus is, perhaps, least accurately determined of any. Some species approach so nearly to Cypræa, others to Helix, to Buccinum, or to Murex, that it is very difficult to trace out the line of demarcation. The most common character of Bulla is an inflated, egg-shaped body; and it will be found to differ from those genera which it most resembles, in some one, if not in more essential points.

To render this genus in some degree more systematical, it is proposed to apportion it into the four divisions, of which the forms are given in the plate of Bulla. Though some, even many of the

species might with great propriety be removed to
other families, still, as the object is not to alter,
but to elucidate, it is trusted that all the shells
which can, consistently with the definition, be included in the genus, and all those which have found
admission rather because they do not belong to any
other, than that they strictly correspond with the
character of this Linnæan tribe, will be arranged
with facility under one or other of the above-men-
tioned sections.　In more than one species of this
genus the common order of nature seems to be reversed : the shell is enclosed within the mantle of
the animal, instead of forming an exterior shield.
So perfectly is it concealed in the B. *aperta*, for
instance, that no inexperienced eye would expect
to find a regular testaceous specimen imbedded in
the unsightly slug.

Many of the Bullæ are river shells; but the marine species are usually imbedded an inch or two
below the surface of the sand.　The name Bulla,
a bubble, is very descriptive of the swelled round
form which characterises the legitimate offspring
of this family, and should exclude those which
have been surreptitiously introduced.

VOLUTA.

A. Aperture entire. (Plate XV. Fig. 1.)
B. Subcylindrical, emarginate. (Fig. 2.)
C. Oval, effuse, emarginate. (Fig. 3.)
D. Fusiform. (Fig. 4.)
E. Ventricose, spire papillary at the tip. (Fig. 5.)

SHELL univalve, convolute. Aperture not elongated, subeffuse. Columella plaited, without either lip or umbilicus.

Although this genus contain many apparently incongruous shells, yet as it stands in the Linnæan arrangement it is more easily discriminated than almost any other, because all shells are referred to it which possess the plaited columella, except a few, which from evident and strong analogy are to be attached to some other genus. The plaits in the columella, which vary in number, but are never very numerous, may be easily distinguished from the teeth of a Cypræa ; these last being uniformly placed on a thickened columellar lip, and having corresponding dentations on the outer margin of the aperture. The plaits are usually longitudinally inclined, and not horizontally, as in the plaited division of the Murices.

It has been objected to this genus, that it contains shells not only differing extremely in form, but inhabited by animals generically distinct. The first objection, if it be valid, would apply to several other genera, confessedly the work of the same description of worm; the second is only an additional proof of a fact universally admitted, that, *if it were possible to obtain it,* a natural system would be better than an artificial one. Until this desirable method of arrangement be perfected, we must condescend to adopt some one or two strong characters of the style of architecture for the classification of testaceous edifices, and leave the fame and family of the architect to the care of future generations.

The cognomen of the genus, signifying " rolled up *cylindrically,*" is justly applied, and sufficiently authorises the separation of CONUS by Linnæus, the shells of the latter being *conically* convolute.

BUCCINUM.

A. Inflated, rounded, thin, subdiaphanous, and brittle. (Plate XVI. Fig. 1.)

B. With a short exserted reflected beak : lip unarmed outwardly. (Fig. 2.)

C. Outer lip prickly on the hinder margin : in other respects as B. (Fig. 3.)

D. Columellar lip dilated and thickened. (Fig. 4.)

E. Columellar lip appearing as if worn flat. (Fig. 5.)

F. Smooth, and not enumerated in the former divisions. (Fig. 6.)

G. Angular, not previously enumerated. (Fig. 7.)

H. Tapering, subulate, smooth. (Fig. 8.)

Shell univalve, spiral, gibbous. Aperture ovate, ending in a canal turned to the right, *i. e.* from the exterior lip, with a short rostrum or beak. Interior lip flattened.

The numerous divisions of this genus include necessarily a great variety of shells, yet all agree-

ing in the short canal turned more or less towards
the right. There are some few species of Buccinum which might, in the formation of a new and
more accurate catalogue than we now possess, be
transferred to Strombus and Murex, especially
from the tapering division, which is far from being
well defined in either of the three genera. The
direction of the canal is, however, a character not
easily mistaken; and if we adhere rigidly to this,
we shall find that there are not a great many species which could be better situated than they are
at present, in one or other of the natural families
united to the genus Buccinum, in consequence of
possessing the peculiar construction of rostrum.

The word Buccinum is derived from βυϰάϻ, *a
trumpet* or *horn;* and was applied by Pliny to a
certain class of shells with a round emarginate
mouth. The genus to which it is now confined is,
perhaps, less generally like a trumpet than many
others.

STROMBUS.

A. Outer lip digitated, projecting in linear
   divisions or claws. (Plate XVII. Fig. 1.)
B. Lobed. (Fig. 2.)
C. Dilated. (Fig. 3.)
D. Tapering, with a long spire. (Fig. 4.)

SHELL univalve, spiral, expanded. Aperture having the lip usually dilated, and ending in a canal
inclined towards the left, or from the columella.

A very remarkable peculiarity in this genus is
the sinus, or impression of the outer lip, which is
situated near the base, but is not at all connected
with the channel of the rostrum. This furnishes
a better discriminative mark than the direction of
the beak, which is often far from evident. It must
however be remarked, that several species, particularly in the last section, do not possess the separate indentation, which are apparently Strombi
from the obliquity of the canal; but as this conformation is not found in any but a Strombus, it is
a sure criterion of the genus, where it does exist.

Perhaps very accurate investigation might induce a corrector of the Linnæan catalogue totally to exclude all shells in which it is not found. More than one species might safely be considered Murices, others Buccina, and others even Helices. Many of the unfinished shells of this genus resemble Coni, and are only to be detected by the length of their spire, which generally exceeds that of any Conus; and sometimes by their tuberculated whorls, which are extremely rare in the latter genus. The animals of the Strombi are little known; but it is evident from the lateral sinus, which unquestionably has not been formed in vain, that there is some specified difference which distinguishes them from their nearest allies. Some peculiar construction requires the provision of an additional duct. The term Strombus was originally used for all turbinated shells, from στροβέω, *to turn round;* but is now exclusively conferred on those ostracodermata which are distinguished by the construction above explained.

A. Spinous, with a produced beak. (Plate XVIII. Fig. 1.)

B. Having longitudinal sutures, or varices, expanding into crisped foliations : beak abbreviated. (Fig. 2.)

C. With thick protuberant, rounded, sutures; varicose. (Fig. 3.)

D. More or less spinous, and without any beak. (Fig. 4.)

E. Unarmed with spines, and having a long, straight, subulate beak. (Fig. 5.)

F. Tapering, with a long spire and short beak. (Fig. 6.)

G. With a toothed columella. (Plate XXIII. Fig. 2.)

SHELL univalve, spiral, often formed with longitudinal membranaceous sutures. Aperture terminating in a canal either straight or turned up backwards, and not inclining either to the right or left.

The very peculiar form of the aperture in this genus is a strong and never-failing distinctive feature. This is oblong-oval, or perfectly oval, seldom ovate, and does not gradually contract into a canal, but abruptly opens into it at the same or nearly the same width which it continues to retain throughout the whole length of the beak. Even in the division which is destitute of an exserted beak, the same contour is no less observable. The straightness of the rostrum is a much more questionable mark, than the outline of the aperture, which when once understood cannot easily be mistaken.

The last section which has been added to those in the Linnæan arrangement of this genus, appears to be absolutely indispensable. A plaited columella having been assumed as the generic character of Voluta, it can only be with some strong reason that shells so furnished are admitted into other genera : such a reason is however found for the plaited Murices in the structure of the mouth, which at once identifies them, notwithstanding any other claim. But they surely deserve to be set apart as a distinct branch of the family : for generic characters, when they lose their validity as such, ought not at once to degenerate into mere specific ones, but should, in subordination, collect the species into natural divisions.

The genus owes its designation to a certain number of its shells, which are *rough* and *rock-like*. The PURPURÆ of the ancients form the second section. From these shells, or rather from their inhabitants, was expressed the famous Tyrian dye, the costly purple which constituted an attribute of imperial dignity. A single vein situated near the head of the fish contains this colouring liquor which was formerly considered so precious, but has of late years yielded its claims on public estimation to other purple dyes equally beautiful and more easily attainable.

The quality is not confined to this one family of Testacea, but has been discovered in certain species of Buccinum, and may very probably belong to some Limaces which inhabit other genera.

TROCHUS.

A. Erect, pillar perforated. (Plate XIX. Fig. 1.)

B. Imperforate, erect; the umbilicus closed. (Fig. 2.)

C. Tapering, with an exserted pillar, and falling on the side when placed upon the base. (Fig. 3.)

SHELL univalve, spiral, subconical. Aperture four-sided, and somewhat angular; or more round, having the upper part of the margin converging towards the pillar. Columella oblique.

Some species, the aperture of which tends to a circular or oval form, are distinguished by a tooth-like projection. There are no two genera more confounded with each other, or more difficult to be discriminated by the inexperienced student, than this and the following one. It is scarcely possible to define the boundary at which the Trochi with rounded apertures are supposed to end, and the Turbines with imperfectly circular mouths to begin their jurisdiction. The true form of the Trochus is that of a pointed cone, capable of standing nearly perpendicularly, or but little inclined, upon the flattened base of the last whorl : the aperture *broader than long*, angular at the lower extremities of the columella, and at the carinate margin of the outer lip, is so situated as to be nearly horizontal when the shell is placed in an upright posture. From this most perfect structure, to the vertical and circular formation, there are so many gradations and varieties of aperture, that it were endless to describe them. The most simple, though not an unexceptionable, rule is, to consider all specimens as belonging to this genus which have any angular tendency in the contour of the mouth, and are, as to their general appearance, *top*-shaped, in conformity with the meaning of their name, derived from the Greek τροχός.

The remarkable faculty which the T. *conchyliophorus* possesses of attaching stones and testaceous fragments to his shell, obviously during the period of its formation, is not easily explained. We must suppose, either that some very strongly adhesive matter is combined with the calcareous secretions, or that the animal is singularly tranquil in his disposition.

H

TURBO.

A. Pillar margin of the aperture dilated, imperforate. (Plate XIX. Fig. 4.)

B. Solid, imperforate. (Fig. 5.)

C. Solid, perforate. (Fig. 6.)

D. Cancellate. (Fig. 7.)

E. Tapering. (Fig. 8.)

SHELL univalve, spiral. Aperture coarctate,
round, entire.

The shells of the genera Turbo and Trochus
are extremely similar, and are much intermixed in
the Linnæan arrangement; yet the round aperture
of the one, and the angular form which charac-
terizes the other, ought sufficiently to point out to
which each belongs. The probability is, however,
that the Turbines which have escaped from their
proper genus, are fewer than the strangers which
have been admitted into it. The rotundity of the
mouth is a less indeterminate diagnosis than that
of almost any other genus; and even in these species in which it has an inclination to oval or ovate,

still it is, or ought to be, without the slightest angularity. Much as the cancellate and tapering sections differ in outward properties of figure from the preceding divisions, yet the internal chamber is equally round and entire in all. There does not appear, therefore, any just reason for separating from the genus those species which are not defective in the principal generic character laid down; but some approach too near the Helix and other tribes, to remain unmolested in their present situation. Doubtful species, of which there are unquestionably but too many, must be learnt by experience only. The exact limits of each genus should be well understood, and then no great inconvenience can arise from the errors of a catalogue.

The name of this genus is as closely connected with that of its predecessor, as are the individuals which severally compose them; Turbo signifying, like Trochus, any thing which *whirls round*, a top. Both genera are found among the rocks on craggy shores, and on the sands, after a storm has detached them from their accustomed refuge.

HELIX.

A. Whorls compressed, or acutely carinate. (Plate XX. Fig. 1.) (Plate XXVI. Fig. 1.)
B. Umbilicate, whorls rounded. (Fig. 2.)
C. Imperforate, whorls rounded. (Fig. 3.)
D. Tapering. (Fig. 4.)
E. Ovate, imperforate. (Fig. 5.)

SHELL univalve, spiral, subdiaphanous, brittle. Aperture coarctate, lunate, or circular, having the segment of another circle taken from the whole area; often obovate.

The whorls are contiguous, and there is not the smallest columellar lip in this genus, but the body *uniformly* projects convexly into the circumference of the aperture. In a single species of division A., the one figured in our plate, the whorls are carinate longitudinally, instead of transversely, so as to form a flattened two-edged shell. This in Gmelin's edition of the "Systema Naturæ" constitutes a section by itself; but as it answers to the terms of the definition of the next division, it is perhaps scarcely necessary. The aperture of many of the species in the first family is so very similar to that of Trochi, that in arranging them we must be careful to observe the convexity of the columella : this sufficiently distinguishes the Helices.

The general character,—which is obvious in almost all the species, whether terrestrial, fluviatile, or marine,—of thinness and transparency, is of great assistance in acquiring a thorough knowledge of the genus. In this property many of the Bullæ partake; but they are not likely to mislead, on account of other generic distinctions. It may be remarked, that the greater number of shells which are not oceanic, are far more fragile and diaphanous, than those which have to endure the rough beating of a boisterous sea. Those which are found in still ponds and muddy ditches are, many of them, scarcely able to resist the slightest pressure.

The name, ἕλιξ, implies merely a shell constructed with a *spire*, or with *circumvolutions* of the whorls. It might therefore include several other genera.

NERITA.

A. Umbilicate. (Plate XX. Fig. 6.)
B. Imperforate, with the lips toothless. (Fig. 7.)
C. Imperforate, with the lips toothed. (Fig. 8.)

SHELL univalve, spiral, gibbous, rather flat underneath. Aperture semiorbicular, or semilunar;
having, uniformly, the pillar lip, or columella
straight.

In no one genus is the diagnosis more perfect
than in Nerita. The third division is totally unlike all others: the flat, toothed, inner lip and
narrow throat, which constitute its very obvious
characters, are not to be found in the slightest degree of similarity in any but this one family. A
few species of the section B. resemble Turbines;
but their columella is always to be distinguished by
its flatness. The umbilicate shells might in some
instances be misplaced in the genus Helix, the
whorls having much the same external figure and
simplicity of colour; the straight pillar lip is, however, in all these individuals sufficiently evident,

and cannot be confounded with the convex elabiate side of the aperture, which is the generic distinction of the tribe of Helices.

Nothing can exceed the beauty and delicacy of the miniature painting with which many of the Neritæ are adorned. When viewed with a magnifying-glass, the most highly finished touches, upon the smallest scale, are discernible on their enamelled surface. The number of species and varieties approximating to each other very closely, or only differing in some one nice point, renders this genus as difficult to be well arranged specifically, as it is easy to be discriminated from other genera.

This difficulty is much increased by the uniformity of shape, especially in the last division; for it becomes necessary, where the outline is nearly similar, to have recourse to the tones of colour for specific marks.

Νηρίτης, from whence this description of shell is called Nerita, may probably be derived from νηρός, hollow, the superior whorls occupying but a small share of the internal cavity.

HALIOTIS.

A. Perforate. (Plate XXI. Fig. 1.)

B. Imperforate. (Fig. 2.)

SHELL ear-shaped, open ; spire lateral, and nearly hidden ; disk, in the first division, longitudinally perforated with pores.

The perforations have been considered as essential to the generic character of Haliotis : but still, with some inconsistency, more than one imperforated species have found admittance ; while others, which in most respects are intimately allied to the genus, have for the same reason been excluded. So precisely similar is the entire specimen figured in our plate to the penetrated one, that no reasonable doubt can exist of their belonging to the same genus. The situation of the spire, the general outline, the involution of the margin, the obvious mode of increment, the internal and external superficies of the shell,—all point out the necessity of a distinct division to comprehend such shells as these, which have no generic deficiency but that of the foraminal ducts.

Did the animals inhabiting the shells of one section differ from that which constructs the other, it would not prevent a generic union of them, according to our system ; but no such fact of their dissimilarity is ascertained. Some of the identical species open, and make use of a different number of their siphons. Why may not then a Limax be permitted to choose a water-tight receptacle, though his fellows prefer one that is pervious to the winds and waves ?

The Haliotis, from ἅλς, *the sea*, and ὦτα, *ears*, is thus denominated on account of its ear-like form. The animal is attached by so adhesive a property to the surface of the rocks, that it requires the utmost force to disengage it, though by a spontaneous action it is able to remove with facility from place to place. It is probable that in some individuals of this family, as in some of Helix and Patella, the shell does not cover the whole body of the Molluscous worm, but merely the vital organs.

A. Furnished with an internal lip or chamber. (Plate XXI. Fig. 3.) (Fig. 4.)

B. Margin angular, or irregularly toothed. (Fig. 5.)

C. With a pointed, recurved apex. (Fig. 6.)

D. Very entire, not pointed at the apex. (Fig. 7.)

E. Having the apex perforated. (Fig. 8.)

SHELL univalve, conical, mostly without spire.

The division C., consisting of shells with a recurved apex, forms a natural link between this genus and the last described. The curvature of the apex approaches more or less to a regular spire, and in some species is precisely of the same description as that of a Haliotis; but, then, the shell is not flat and ear-shaped, and therefore cannot belong to Haliotis, or to any existing genus but Patella, for there are none equally patulous with these two. The gradations in the scale of nature are in general so regular and yet so small, that it becomes no easy task to trace the separating line. Class is linked to class by an order trespassing on both; order to order by an intermediate genus; a doubtful species unites two genera; and varieties confound the limits of a species. This curious course is well illustrated by the observations we may make on the genus Patella, placed as it were on the boundary between those shells which have a regular spire, and those which have it not. At least we may boast of somewhat more accurate ideas upon the subject than those naturalists possessed, who, about two centuries ago, placed Patellæ among the *Bivalves*, because the stone to which the animal adhered served for a second valve.

Some few of the chambered section possess a slight resemblance to Neritæ; but upon close examination, it is found that the margin rises above the flattened dissepiment on all sides, which is therefore not to be compared with the columellar lip of the Nerita. The remaining divisions are of the simplest form; and, as their name PATELLÆ signifies, assume the shape of various *little dishes*, affixed by their tenants firmly to the rocks; with the apex uppermost.

DENTALIUM.

(Plate XXII. Fig. 1.)

SHELL univalve, straight, or nearly so, subconical, tubular, not chambered, open at both ends.

The simple construction of the shells in this genus, and the paucity of species, render it unnecessary to offer any extended remarks on their distinctive character. They are all more or less, as their name expresses, like *teeth* or *tusks*, and are completely separated from the other tubular families, by being entirely without contortion, though somewhat curved.

They are usually discovered partly buried in the sand; and the animal, which has by some naturalists been supposed perfectly free and unattached to his habitation, is seen to shrink deeply into it for protection from impending danger. The opinion, that the Terebella possesses the power of disengaging himself from his shell, has arisen probably from the circumstance of there being no apex, hinge, or visible depression to which the connecting muscle, as in other genera, is attached. It is true, that both in this and the following genus, the animal does not appear to be restrained in his motions within the limits of his house; but it by no means follows that he has not some tie to prevent his leaving it. Were he perfectly at liberty, the form of his receptacle would scarcely preclude the action of the water or accidental violence from forcing him from it against his will.

SERPULA.

A. Flexuous, irregular, and adhering. (Plate XXII. Fig. 2.)

B. Assuming a certain form, detached. (Fig. 3.)

SHELL univalve, tubular, gradually tapering, often interrupted by imperforate partitions at irregular distances. Frequently closed at one end.

By constituting two divisions of this genus, and omitting the character of *adherence* in the generic description of Linnæus, we legally include a great variety of shells, which, without this alteration, were inconsistent with the definition. Many species which should rank in the first division have a great appearance of regularity in their structure, and yet, strictly speaking, are irregular. The end is twisted, sometimes the whole shell, into a spiral form, much in the same manner in all the individuals of the species; but there is no fixed number of circumvolutions, nor any symptom of the animal being directed by an invariable law, as in turbinated and other genera.

The concamerated varieties which are found among the serpulæ do not deserve a separate division, because the conformation appears to originate not in any specific difference, but in an instinctive desire of the animal, for some important purpose, to increase the length of the shell without materially adding to the size of his apartment. The chambers are not connected with each other by any siphunculus or opening; the last can therefore only be inhabited. No sign of the internal dissepiments is visible externally, except in one species, S. *Polythalamia.*

In this genus, as well as in the preceding one, . the animal has been supposed to live with little or no adhesion to his shell; but the degree and mode of his attachment are doubtless as various as the peculiar specific forms.

The *creeping,* tortuous character of the first division of Serpula furnishes us with a clue to the etymon of the generic name, to ἕρπω, Serpo.

TEREDO.

(Plate XXII. Fig. 4.)

SHELL univalve, tubular, tapering, flexuous, penetrating wood. One end is closed by two hemispherical, and the other by two lanceolate valves.

It has been doubted whether the Teredines ought not to be considered as multivalve shells, rather than among the most simple of the univalves. The small valves which are attached in pairs to the fore and hind part of the animal are purely testaceous, and as necessary to the habits of the worm as the tube in which he dwells. The anterior hemispherical valves are placed at an angle, and furnished internally with a long flat and curved tooth, probably intended to strengthen the molluscous head on which this curious boring instrument is fixed. The smaller end of the tube, in which the lanceolate pair are situated, remains at the surface of the perforated wood, and the little valves are used as flood-gates to admit more or less of the water according to necessity. There appears however to be no more reason for classing these shells with the multivalves, than Turbines and others, which possess an operculum, with the bivalves. The only difference between the lid of the Teredo and that of a Turbo is, that the one is constructed of four pieces, the other of a single disk. These pieces cannot constitute *the shell*, because the animal cannot be said to *inhabit* them, as it does the testaceous tube. Neither is there any analogy between them and the accessary valves of Pholas.

At first sight, the Teredo may easily be mistaken for a Serpula, and in many cabinet specimens the valves are lost. There are, however, but three species of the former genus at present known, and their external characters are soon to be distinguished. The first, T. *navalis*, which is the most common, is much more thin and brittle than Serpulæ in general, especially towards the smaller end.

The name Teredo, or τερηδὼν, is derived from τερέω, *to bore*, and is sufficiently descriptive of the mode in which all the species effect their settlement.

SABELLA.

(Plate XXII. Fig. 5.)

SHELL tubular, formed of sandy and calcareous particles agglutinated and inserted in a membranaceous sheath.

. The best claim which the Sabella could assert to be ranked among Testacea is, that the cement with which the fragments of organised and siliceous matter are fastened to the mould is really calcareous, and not an animal gluten only. By immersing a portion of the shell in muriatic acid the calcareous particles will be dissolved, and many, if not all the indissoluble fragments will be precipitated, leaving the membranaceous sheath entire, and impressed with the forms of the substances which were attached to its exterior surface. Still, some will be found to adhere and resist the action of the acid : it is therefore difficult to affirm whether they have been fastened by a gelatinous secretion, or by a calcareous one. In either case, the fibre of the animal membrane growing round a slight projection in the siliceous pebble might be a sufficient bond.

The Trochus *conchyliophorus* derives the same sort of protection from foreign substances, being covered however with entire stones and shells, instead of pulverised or broken pieces of them. In this case no one doubts the fact of its being a true Trochus, because the foundation of the aggregate is regularly spiral, and possesses the characteristic aperture. Why, then, should Sabella be expelled the order, as it often has been, because the sheath is not spiral, but straight and tubular? Some few species may have found admittance, which do not perhaps belong even to the class of Vermes, as the larvæ of some insects are known to provide themselves with a very similar receptacle to that of a real Sabella. There is one never-failing mark of genuineness in most perfect specimens of this genus,—that of the end being, as it were, fringed, or produced into numerous ragged processes of the same nature with the shell. They are, however, so extremely friable, that perfect specimens are rather rare. The denomination of the tribe is taken from the principal constituent of their edifices, Sabulum, *fine gravel* or *sand*.

IT may possibly be remarked, that in the preceding description of the genera of shells, no mention has been made by name, of the systems and opinions of either former or contemporary writers on the subject. This omission has not arisen from any want of respect to the literary labours of those who, with great research and skill, have brought the science of Conchology to its present state, or of those who, like the author, have availed themselves largely of the written documents furnished by works in the German, French, and Latin anguages, which have not been translated into our own. The names of Testaceological writers, and the titles of their books, with short notices on some, will be given in another place : but it has appeared to be one great source of the uncertainty and confusion attending this branch of study, that the beginner had not only to learn a system, but to select one from many extremely different, placed for comparison in the same page. It would be inconsistent with the professed object of this elementary treatise to insert a variety of theoretical matter, which could only tend to divert the mind from the simplicity of the Linnæan system ; it is that alone which the author wishes to illustrate, availing himself of the substance of what has hitherto been published, where it is well authenticated, and not inseparably connected with any other method of arrangement.

It is recommended to the young student in Conchology, to compare carefully the following Specific Descriptions with the Plates to which they refer, at the same time turning to the Nomenclature for the exact signification of the terms. By repeating this exercise till the species adopted as examples of the different genera and sections be perfectly known, a solid foundation will be laid for a more extensive acquaintance with the numerous beautiful and interesting individuals which are all to be arranged under some one or other of these generic forms. Those who may wish to acquire an accurate knowledge of the Linnæan system with still greater certainty, will find much benefit in selecting for examination a good specimen of each shell here illustrated, and placing them in a cabinet in their proper order. A regular series upon this plan, consisting of more or less valuable shells, with references to the annexed drawings, may be procured from Mrs. Mawe, 149, Strand, in whose superb collection almost every instructive subject may be viewed, and from whose liberality and science much information will be derived.

SPECIFIC DESCRIPTION.

PLATE I.

Explanatory of the parts of Univalve Shells.

PLATE II.

Explanatory of the parts of Bivalves and Multi-valves.

PLATE III.

Fig. 1. CHITON *squamosus.*

Shell 8-valved, semistriate ; marginal membrane scaly.

Born. Mus. Cæs. p. 5. Tab. 1. Fig. 1, 2.

Specimen olive without, blue green within. Middle valves divided on each side of the central ridge into two parts, consisting of a triangular compartment striated from the apex to the base, and another very finely marked in the longitudinal direction of the shell; terminal valves lunulate and striate; the scales on the marginal skin in alternate black and white divisions.

Inhabits New South Wales, the Indian and American seas.

———

Fig. 2. LEPAS *Tintinnabulum.*
Shell conic, obtuse, rugged.
Lister Conch. Tab. 443. Fig. 285.
Specimen with six erect valves, reddish purple, rayed with white, striated longitudinally, deeply towards the base; intermediate spaces purplish white, depressed, finely striated transversely. Aperture triangular. Operculum of four valves, two larger.

Inhabits Sumatra.

———

Fig. 3.   L. *anatifera.*
Shell compressed, 5-valved, smooth, placed on a peduncle.
Lister Conch. Tab. 439. Fig. 288.
Specimen white, with a pearly lustre;—Peduncle coriaceous, red or brown, wrinkled towards the

shell, paler and pellucid towards the base; valves finely striate.

Two larger valves three-sided, curvilinear; two upper ones nearly triangular; connecting one curved, narrow, rounded on the back.

Inhabits nearly all seas.

The specimen from which the annexed drawing was made, being a dried one, the peduncle is not inflated, nor does it preserve its original colour; but the skin which adheres to the margin of the valves is of a reddish orange.

———

Fig. 4. PHOLAS *candida*.

Shell oblong, muricate on all sides with decussate striæ.

Lister Conch. Tab. 435. Fig. 278.

Specimen white, very thin, within smooth, and rather silvery, both ends rounded; tooth of the hinge long, thin, and curved; margin of the hinge turgid, and projecting with a sharp curved fold towards the anterior part. One accessory valve lanceolate.

Inhabits the English coast.

PLATE IV.

Fig. 1. HINGE of MYA *truncata.*

———

Fig. 2. MYA *truncata.*

Shell ovate, truncate on the posterior margin, tooth of the hinge projecting forwards and very blunt.

Lister Conch. Tab. 428. Fig. 269.

Specimen dirty white, clothed with a yellowish brown epidermis, which extends beyond the truncate side; thick, convex, deeply marked with irregular, transverse striæ, gaping widely at the truncated end, very smooth within. Solitary broad tooth of the hinge connected with the opposite valve by an interior cartilage.

Inhabits the English coast.

———

Fig. 3. HINGE of SOLEN *Vagina.*

———

Fig. 4. SOLEN *Ensis.*

Shell linear, rather curved; hinge with two teeth in one valve, and one locking between them in the other.

Lister Conch. Tab. 411. Fig. 257.

Specimen white, marked with bluish flesh-coloured streaks and spots, striated transversely in such a manner as that towards the anterior side the shell appears divided into two triangular compartments following the curvature of the lower margin; covered with a yellowish olive transparent epidermis.

Inhabits the English coast.

Fig. 5. S. *radiatus*.

Shell oblong oval, straight, smooth, with an internal rib from the hinge to the base.

Lister Conch. Tab. 422. Fig. 266.

Specimen violet, with four white rays, extremely thin and brittle, rib white, strong; hinge in both valves callous; teeth in both valves bifid.

Inhabits the East Indies.

PLATE V.

Fig. 1. HINGE of TELLINA *radiata*.

Fig. 2. TELLINA *rugosa*.

Shell ovate, with transverse undulated wrinkles;

hinge with two lateral teeth; one primary tooth in the right valve, and two in the left, bifid.

Born. Mus. Cæs. Tab. 2. Fig. 3, 4.

Specimen white, yellowish towards the beaks, smooth within.

Inhabits the Indian and American seas.

———

Fig. 3. T. *planata.*

Shell ovate, compressed, transversely substriate, smooth, margin acute.

Born. Mus. Cæs. Tab. 2. Fig. 9.

Specimen purplish red, with paler concentric bands, and a tinge of yellow about the beaks; thin, pellucid.

Inhabits the English coast.

———

Fig. 4. T. *cornea.*

Shell orbicular, smooth, horn-colour, transversely striate.

Lister Conch. Tab. 159. Fig. 14.

Specimen within dirty white passing into cinereous; without bluish white, polished; transverse striæ unequally deep; covered with an olive brown epidermis, which is darker in transverse bands, one usually darker than the rest. Lateral teeth elongated, inserted; primary ones small.

Inhabits the British coast.

Fig. 5. HINGE of CARDIUM *echinatum*.

Fig. 6. CARDIUM *edule*.

Shell antiquated, with 20—30 rounded ribs, obsoletely imbricate.

Lister Conch. T. 333. F. 170. T. 334. F. 171.

Specimen pale brown, cinereous towards the margin; a dark brown and tawny spot upon the internal cavity of the anterior slope.

This shell, which is the common cockle of the market, is found in great numbers on the British coast, buried in the sand at no great distance from the surface.

Fig. 7. C. *Cardissa*.

Shell cordate, valves compressed, dentato-carinate; beaks approximate.

Lister Conch. Tab. 318.

Specimen white; equivalve; ribs oblique; beaks slightly crossing each other; cordate impression on the area well defined; anterior side nearly flat, posterior one convex except towards the margin.

Inhabits the Indian ocean.

———◆———

PLATE VI.

Fig. 1. HINGE of MACTRA *lutraria.*

———

Fig. 2. MACTRA *Stultorum.*

Shell subdiaphanous, smooth, obsoletely radiate, purplish within; area gibbous.

Lister Conch. Tab. 251.

Specimen pale brown inclining to cinereous, with paler rays, finely striate transversely.

Inhabits the English coast.

———

Fig. 3. HINGE of DONAX *Scortum.*

———

Fig. 4. DONAX *denticulata.*

Shell very obtuse on the anterior side; lips transversely wrinkled; margin denticulate; shell striate longitudinally.

Lister Conch. Tab. 376. Fig. 218, 219.

Specimen white, with obsolete purple rays; area cordate; fore part strongly marked with decussate striæ.

Inhabits the European and American seas.

————◆————

PLATE VII.

Fig. 1. HINGE of VENUS *concentrica.*

————

Fig. 2. VENUS *Paphia.*

Shell subcordate, with thickened transverse wrinkles which are attenuated towards the anterior slope; lips complicated.

Lister Conch. Tab. 279.

Specimen white, with interrupted brown rays, spots and lines; areola cordate, brown; transverse ribs broad and convex.

It has been considered a needless increase of sections to separate, as in Gmelin's edition of Linnæus's "Systema," those shells which are said to be muricate before, from those which are not so, possessing a general subcordate form. There is in fact but one muricated species, V. *Dione*; V. *Paphia* is included in the same division, which has no more title to the distinction than many individuals decidedly " unarmed;" the margin of the area, as is evident in the figure, being wrinkled and not muricate.

Fig. 3. V. *edentula.*

Shell subglobular, lenticular, transversely striate, without teeth.

Lister Conch. Tab. 260. Fig. 96.

Specimen diaphanous, white, when recent with a tinge of red, inside golden, posterior slope ovate, very acute at the margin, area straight, beaks turned towards the areola.

Inhabits the American ocean.

Fig. 4. V. *litterata.*

Shell ovate, subangular before; transversely and longitudinally striate, undulate and rough towards the slopes.

Lister Conch. Tab. 402. Fig. 246.

Specimen brownish white, with brown angular line and characters in rays; the cavity of the ligament long and rather broad; beaks extremely small.

Inhabits Europe, and more rarely India.

Fig. 5. HINGE of the flatter Valve of SPON-
DYLUS *Gædaropus.*

———

Fig. 6. SPONDYLUS *Gædaropus.*
Shell slightly eared, spinous.
Lister Conch. Tab. 206. Fig. 40.
Specimen white, orange towards the margin,
pectinate with knotted ribs on the upper valve,
lower one tuberculate and. rough, obliquely pectinated.

Inhabits the Mediterranean, Red, Indian, and
American Seas.

The varieties of this species are extremely numerous; those which are furnished with spines are
considered the most valuable, and increase in estimation in proportion to the length and beauty of
the appendages.

———

PLATE VIII.

Fig. 1. HINGE of CHAMA *Gigas.*

———

Fig. 2. CHAMA *Gigas.*
Shell folded, ribs with arched scales; posterior
slope gaping.

Lister Conch. Tab. 351. Fig. 189.

Specimen clear white; gape of the posterior slope lanceolate, the circumference of which is tumid and crenate; margin of the shell deeply crenate; hinge furnished with an anterior tooth besides the usual callus.

Inhabits the Indian Ocean.

Fig. 3.   Hinge of Arca *granosa.*

Fig. 4.   Arca *tortuosa.*

Shell parallelopiped, striate; larger valve obliquely carinate.

D'Argenville Conch. Tab. 19. Fig. 1.

Specimen white; striated longitudinally and transversely; valves dissimilar, much twisted; margin subcrenate.

Inhabits the Indian Ocean.

Fig. 5.   A. *Noæ.*

Shell oblong, striated, emarginate at the apex, beaks remote, margin gaping.

Lister Conch. Tab. 368. Fig. 208.

Specimen white and brown, rhomboidal; external margin crenate, internal one entire; area with angular brown lines and impressed marks.

Inhabits the Mediterranean, Red, Atlantic, and Indian Seas.

———

Fig. 6. A. *granosa*.

Shell subcordate with muricated ribs.

Lister Conch. Tab. 241. Fig. 78.

Specimen white, nearly equilateral; ribs 20, tuberculate with obtuse spines.

Inhabits the Indian and American Ocean.

———

Fig. 7. A. *Glycymeris*.

Shell suborbicular, gibbous, substriate.

Lister Conch. Tab. 247. Fig. 82.

Specimen white, marked with irregular, interrupted, transverse bands, and angular lines of reddish brown, white within; the longitudinal and transverse striæ extremely fine; hinge bowed.

Inhabits the English and other seas.

———

PLATE IX.

Fig. 1. OSTREA *Radula*.

Shell sub-equivalve, with 12 convex rays, and decussate, crenate striæ.

Lister Conch. Tab. 175. Fig. 12.

Specimen twice the size of the figure, white spotted with reddish brown; ears rough, obliquely wrinkled.

Inhabits the Indian Ocean.

———

Fig. 2.   O. *varia.*

Shell equivalve; about 30 compressed rough and prickly rays; one ear very small.

Lister Conch. Tab. 178. Fig. 15.

Specimen yellow brown with darker clouds; rather convex; rays obsoletely spinous, intermediate spaces not striate, the smaller ear spinous, the larger with wrinkled plaits and 5 spines beneath.

Inhabits the English coast.

———

Fig. 3.   O. *fasciata.*

Shell equivalve; 20 rough rays; intermediate furrows striate; ears equal, very small.

Lister Conch. Tab. 177. Fig. 14.

Specimen white, pellucid gaping on both sides, within finely striate; margin crenate.

Inhabits the Atlantic Ocean.

Fig. 4. O. *Folium.*

Shell ovate, obtusely plaited at the sides, parasitical.

D'Argenville Conch. Tab. 19. F.

Specimen pale purplish brown, underneath inclining to white; hinge with a triangular sinus; upper valve turgid towards the middle, rough and transversely ribbed on both sides; lower valve smaller, flat, and channelled through the middle.

Inhabits the Indian Ocean.

———

Fig. 5. O. *isogonum.*

Shell equivalve, lobed, the larger lobe at right angles to the hinge.

Rumph. Mus. Tab. 47. I.

Specimen inclining to violet, with shades of black, within pearly; lamellate; beak unclosed.

Inhabits the Indian Ocean and South Seas.

———◆———

PLATE X.

Fig. 1. ANOMIA *vitrea.*

Shell nearly orbicular, ventricose, hyaline, extremely thin; lower valve with two bony rays at the hinge, besides lateral teeth; upper valve with a prominent perforated apex.

Born. Mus. Cæs. p. 116. *vign.*

Specimen very pale cinereous; striated transversely; striæ scarcely perceptible. The annexed figure represents the interior of the upper valve with the lateral teeth of the hinge.

Inhabits the Mediterranean Sea.

———

Fig. 2.   A. *Ephippium.*

Shell suborbicular, rough, undulate, the flatter valve perforated.

Lister Conch. Tab. 204. Fig. 38.

Specimen white, convex valve purplish, saturated towards the apex, silvery within and without; translucent, brittle; rather lamellate.

Inhabits the Indian and American Ocean.

———

Fig. 3.   Hinge of A. *Sella.*

———

Fig. 4. MYTILUS *Frons.*

Shell plaited; one lip scabrous.

Born. Mus. Cæs. p. 121. *vign.* Fig. 6.

Specimen reddish, oval; the margin on both

sides plaited, denticulate; lower valve smooth, channelled in the middle, upper one rough with elevated points upon the margin, and an elevated central rib.

Inhabits the American Ocean.

———

Fig. 5. M. *margaritiferus*.

Shell compressed, flat, suborbicular; hinge transverse, imbricate, having the laminæ toothed in rays.

Lister Conch. Tab. 223. Fig. 57.

Specimen whitish with crimson rays, within pearly; being young, it does not possess the tooth-like scales which are found in older shells. Hinge very straight, the length of the whole shell.

Inhabits the American Ocean.

A distinct variety, differing in size and in other respects from the East Indian Pearl-bearing muscle.

———

Fig. 6. M. *edulis*.

Shell smooth, violet; valves rather carinate before, retuse behind, beaks pointed.

Lister Conch. Tab. 362. Fig. 200.

Specimen deep violet; beaks white; somewhat triangular.

Inhabits most seas, at the bottom of which it is found adhering by the byssus in large clusters. It is generally known as the common muscle.

———◆———

PLATE XI.

Fig. 1.   PINNA *pectinata*.

Shell transversely striate, wrinkled towards the base.

Gualt. Test. Tab. 79. A.

Specimen horn-colour with darker clouds; about 10 obsolete ribs converging to the smaller end; triangular; hyaline; brittle.   A smooth variety.

Inhabits India and Europe.

———◆———

PLATE XII.

Fig. 1.   ARGONAUTA *Argo*.

Shell having the keel slightly toothed on both sides.

Lister Conch. Tab. 254, 255.

Specimen white, with undulated, smooth, and bifurcated ribs; teeth of the keel brown towards the apex, extremely thin and brittle.

Inhabits the Mediterranean and Indian Ocean.

———

Fig. 2.   NAUTILUS *Pompilius.*

Shell with a cordate aperture, whorls obtuse and smooth.

Lister Conch. Tab. 550. Fig. 1, 3.

Specimen white, with flexuous yellowish tawny rays and streaks : the inner whorls dark brown ; interior surface fine mother-of-pearl; umbilicus pervious.

Inhabits the Indian and African Ocean.

———

Fig. 3.   N. *Spirula.*

Shell with an orbicular aperture; whorls cylindrical, remote.

Lister Conch. Tab. 550. Fig. 2.

Specimen white, within pearly; whorls gradually decreasing towards the point of the spire; the last whorl excentric, continued, in a perfect shell, into a long cylindrical straight tube; siphunculus lateral.

Inhabits the American and Indian Ocean.

PLATE XIII.

Fig. 1. CONUS *Virgo.*

Shell conical, with a bluish base.

Lister Conch. Tab. 754. Fig. 2.

Specimen pale yellow, base purplish blue, smooth.

Inhabits the African Ocean.

———

Fig. C. *Ebræus.*

Shell ovate, white, with transverse bands formed of black spots.

Lister Conch. Tab. 779. Fig. 25.

Specimen reddish, parallelogramic spots purplish black.

Inhabits India.

———

Fig. 3. C. *Textile.*

Specimen marked with yellow reticulated veins, with yellow and brown spots.

Lister Conch. Tab. 788. Fig. 40.

Shell white, with three interrupted orange bands.

Inhabits Asia.

———

Fig. 4. C. *Tulipa.*

Shell oblong, gibbous, smooth, aperture spreading.

Lister Conch. Tab. 764. Fig. 13.

Specimen white, with bluish, red, and yellow clouds, and transverse interrupted chesnut lines; aperture bluish, base marked with obsolete oblique striæ; spire acute, smooth, spotted, finely striated transversely.

Inhabits India, Africa, and South America.

———◆———

PLATE XIV.

Fig. 1. CYPRÆA *Arabica.*

Shell subturbinate, marked with Oriental characters; longitudinal stripe simple.

Lister Conch. Tab. 659.

Specimen above with bluish-white spots, between them tawny; confluent brown and purple spots on the thickened sides, within pale violet, lips somewhat rounded; teeth brown.

This second variety of the C. *Arabica* is so very different from the first, that it might without impropriety constitute a separate species. The form of the first is longer and narrower, more lengthened at the base, depressed and not elevated upon the back; lips perfectly flat, or slightly concave and immaculate, with pale chesnut teeth ; in the second the lips are rather convex, and on the columella there is a large brown, tawny, or cinereous spot ; teeth dark brown. The characteristic marking is much more distinct in the former than the latter, and less interrupted by reticulated lines.

Inhabits India.

———

Fig. 2.   C. *Caput serpentis*.

Shell triangularly gibbous, rather obtuse at the base.

Lister Conch. Tab. 702. Fig. 50.

Specimen brown, spotted with white on the upper part, a pale fulvous spot at each extremity, beneath white.

Inhabits the Mauritius.

———

Fig. 3.   C. *lurida*.

Shell rather turbinate, lurid, slightly banded; extremities yellow, with two black spots.

Lister Conch. Tab. 671. Fig. 17.

Specimen beneath white; above obscurely marked with two pale cinereous bands.

Inhabits the Mediterranean and Atlantic.

This species stands in the first section of the

genus in Gmelin's edition; but is here selected as an example of the third, being much more remarkably umbilicate than many which are so denominated.

Fig. 4.   C. *Moneta.*

Shell margined, knotty, white.

Lister Conch. Tab. 709. Fig. 59.

Specimen on the back purplish white, outer lip with a row of raised knots, and a single one near the anterior part of the inner lip.

Fig. 5.   BULLA *Volva.*

Shell birostrate ; beaks elongated, acute, and striate.

Lister Conch. Tab. 711. Fig. 63.

Specimen white with a tinge of pink within, thin, subglobular, finely striated transversely ; beaks rather bent backwards.

Inhabits Jamaica.

Fig. 6.   B. *Ficus.*

Shell obovate, club-shaped, reticulately striate; beak exserted; suture of the spire obliterated.

Lister Conch. Tab. 751. Fig. 46. a.

Specimen white, with purplish brown points and spots, and five whitish transverse bands irregularly spotted; interior surface polished, lilac; aperture broad; beak open, hollow.

Inhabits the American and Indian Ocean.

This species resembles a Murex, but the canal is wide, and not contracted at its entrance to the aperture, the latter gradually, and not, as in a true Murex, suddenly opening into the beak; the outer lip is nearly straight, and not compressed towards the columella at the lower part of the body. This shell cannot therefore be considered as belonging to Murex, or any other than the genus Bulla;— no other, besides these two, possessing an elongated straight beak.

Fig. 7.   B. *Ampulla.*

Shell rounded, opaque; vertex umbilicate.

Lister Conch. Tab. 713. Fig. 69.

Specimen pale reddish brown, mottled with cinereous spots, and marked with two darker bands; columella covered with the inner lip, white.

Inhabits all seas.

The variety to which the figured specimen

belongs differs uniformly from that which is without bands, in having the outer lip rounder and more expanded.

Fig. 8. B. *Terebellum.*

Shell cylindrical; spire subulate; base truncate.

Lister Conch. Tab. 736. Fig. 30, 31.

Specimen white, marked with ochraceous transverse and zigzag lines.

Inhabits the Indian Ocean.

PLATE XV.

Fig. 1. VOLUTA *Auris Midæ.*

Shell coarctate, oblong-oval; spire rough; columella two-toothed.

Lister Conch. Tab. 1058. Fig. 6.

Specimen brown, solid, longitudinally wrinkled and transversely striate; aperture white, waxy, long, wider towards the base. Six whorls.

Inhabits India.

Fig. 2. V. *Oliva.*

Shell smooth; columella obliquely striated.

Lister Conch. Tab. 719. Fig. 3. Q.

Specimen white, with very pale reddish brown angular streaks and two obsolete bands.

Inhabits the Indian Ocean.

The varieties of this species are so numerous, and many of them so nearly allied to *Ispidula* and *Utriculus*, that it is scarcely possible to distinguish them. The characters of the three species are ill defined, and applied indiscriminately to each. It were perhaps better to confine the shells with a flattened spire to *Oliva*, those with a very long one to *Utriculus*, and the intermediate degrees to *Ispidula*. The annexed specimen would then be a variety of *Ispidula*.

Fig. 3.   V. *Persicula*.

Shell smooth; spire retuse, umbilicate; columella with seven plaits; margin of the outer lip crenate.

Lister Conch. Tab. 803. Fig. 9.   •

Specimen white, with chesnut transverse lines.

Inhabits the African Sea and Senegal.

Fig. 4.   V. *Papalis*.

Shell emarginate, transversely striate; margin of the whorls and outer lip denticulate; columella with four plaits.

Lister Conch. Tab. 840. Fig. 68.

Specimen white, with large bright yellowish red spots. This second variety differs in many respects materially from the first. It is always much smaller, seldom exceeding the size of the specimen, whereas the other is usually from three to six inches . in length; the denticulate margin of the whorls is not compressed towards the suture; the transverse triple row of impressed points is more distinct, and the holes are deeper on the body than on the superior part of the spire; the red spots are much larger in proportion to the shell, and paler. The outer lip is not denticulate.

Inhabits the Indian Ocean.

———

Fig. 5. **V.** *Glans.*

Shell cylindrical; columella with three teeth, emarginate; aperture effuse, spreading

Chemn. Conch. 10. Tab. 148. Fig. 1393, 1394.

Specimen yellowish brown, suture of the whorls obsolete.

Inhabits the Eastern shores of Africa.

PLATE XVI.

Fig. 1. Buccinum *Dolium.*

Shell ovate, surrounded with transverse remote flattened ribs; beak rather prominent.

Lister Conch. Tab. 899.

Shell white, with square ochraceous spots on the raised belts; very thin; columella twisted.

Inhabits Sicily, Africa, and India.

———

Fig. 2. B. *Areola.*

Shell substriate; belted with four rows of quadrilateral spots; aperture toothed; beak recurved.

Lister Conch. Tab. 1012.

Specimen white, with a brownish tinge; mouth white; spots pale brown; shell transversely striate; inner margin of the outer lip toothed.

Inhabits the Mediterranean and Indian Seas.

———

Fig. 3. B. *Vibex.*

Shell entirely smooth, slightly folded, and crowned with obsolete tubercles.

Lister Conch. Tab. 1015. Fig. 73.

Specimen purplish white, marked with two indistinct pale brown bands, and longitudinal waved streaks, somewhat tessellated with obsolete lines and brown dots at the points of section; lengthened spots on the outer lip dark brown.

Inhabits the American, Indian, and Mediterranean Seas.

The Linnæan species *Erinaceus* and *Vibex* are so obviously the same, and pass so imperceptibly from the plaited and crowned to the even surface, that the characters of both have been here united under the title *Vibex*, as most appropriate. The figured specimen is an exact link between them, the upper whorls and great part of the inferior one being perfectly smooth, the plaits and papillæ extending but a very little distance from the outer lip. In plate XXIV. the young shell is given, which by its marking is an *Erinaceus;* by its form, a *Vibex*. The peculiar black spot at the extremity of the Cauda is a sufficient proof of their identity.

——

Fig. 4.   B. *Pullus.*
Shell gibbous, obliquely striated.
Lister Conch. Tab. 971. Fig. 26.

Specimen cinereous, with one whitish band; lips white and highly polished; outer lip toothed internally; obliquely plaited.

Inhabits the European Seas.

Fig. 5. B. *patulum*.

Shell muricate; outer lip crenate; columella flattened, rather concave, sickle-shaped.

Lister Conch. Tab. 989. Fig. 49.

Specimen white, with dark brown transverse striæ, lines, and tubercles; the brown striæ more or less confluent in parts, so as to form longitudinal dark rays and transverse belts; within bluish, and shewing the external white belts through the substance of the shell; columellar lip bright orange.

Inhabits Africa and America.

Fig. 6. B. *spiratum*.

Shell smooth; whorls separated by a broad canal; columella abrupt, perforated.

Lister Conch. Tab. 981. Fig. 41.

Specimen white, with ochraceous spots; aperture ovate, emarginate at the top; umbilicus usually very broad and deeply perforated, but in this variety nearly closed, leaving but a very small hole visible; columellar lip white and polished.

Inhabits India and China.

———

Fig. 7.   B. *reticulatum.*

Shell ovate, transversely striate, and longitudinally ribbed; aperture toothed.

Lister Conch. Tab. 966. Fig. 21.

Specimen brownish white inclining to cinereous, with a dark reddish brown line round the upper part of the whorls; lips polished, white.

Inhabits the Mediterranean, English, and Æthiopic Seas.

———

Fig. 8.   B. *strigilatum.*

Shell tapering; whorls bifid, obliquely striate.

Lister Conch. Tab. 845. Fig. 73.

Specimen dirty white, with pale reddish brown irregular marks; longitudinal striæ composed of fine impressed dots.

Inhabits Asia.

PLATE XVII.

Fig. 1. STROMBUS *Pes Pelecani.*

Shell having the outer lip palmate, with four angular and flattened processes; aperture smooth.

Lister Conch. Tab. 865. Fig. 20.

Specimen white, within glossy, with three rows of tubercles on the whorls.

Inhabits nearly all seas.

———

Fig. 2. S. *Auris Dianæ.*

Shell having the outer lip mucronate; back muricate; tail erect, acute.

Lister Conch. Tab. 872. Fig. 78.

Specimen white; back pale brown with whitish spots, towards the extremities and margin purplish; interior of the outer lip, flesh-colour; columella white, crowned with two rows of tubercles.

Inhabits Asia.

———

Fig. 3. S. *Urceus.*

Shell having the outer lip attenuated, retuse, short, and striated; body and spire with tuberculated plaits; aperture two-lipped, unarmed.

Lister Conch. Tab. 857. Fig. 13.

Specimen white, with livid spots, forming two obsolete bands; inner lip ochraceous with darker striæ, the striæ of the outer lip brownish purple; tubercles on the lower whorl flat on one side.

Inhabits the Indian Ocean.

Fig. 4. S. *palustris.*

Shell rather smooth; outer lip disengaged at the base.

Lister Conch. Tab. 837. Fig. 63.

Specimen fuscous, thick; lower whorl twice as large as the next, the others longitudinally plaited and transversely striate, with three impressed lines.

Inhabits the Indian Ocean.

PLATE XVIII.

Fig. 1. MUREX *Tribulus.*

Shell ovate, with three rows of setaceous spines, striate; beak elongated, subulate, straight, and spined in the same manner as the body.

Lister Conch. Tab. 902. Fig. 22.

Specimen pale yellowish brown, with brown

spots on the transverse raised lines and striæ; aperture white.

Inhabits Asia, America, and the Red Sea.

Fig. 2. M. *ramosus*.

Shell with a triple row of foliations; spire contiguous; beak truncate.

Lister Conch. Tab. 946. Fig. 41.

Specimen white, immaculate, transversely striate.

Inhabits Asia, America, Persia, and the Red Sea.

Fig. 3. M. *Rana*.

Shell rough, having two opposite compressed varices; belts muricate; aperture ovate and rather toothed.

Lister Conch. Tab. 995. Fig. 58.

Specimen reddish brown inclining to cinereous, and much paler towards the varices; two muricate belts on the body; transverse striæ raised and granulate.

Inhabits Asia.

The second variety of this species differs so much from the present one, that it cannot properly be admitted into the same section. The varices are spined, the shell is longer, narrower, and more flattened; the striæ little raised, and but few of them granulate. They resemble each other most in colour.

Fig. 4. M. *Mancinella.*

Shell ovate, with obsolete spines; transversely striate; aperture toothless.

Lister Conch. Tab. 957. Fig. 10.

Specimen yellowish white, with longitudinal brown rays and spinous protuberances; aperture white, with a tinge of pale orange; four raised dots on the interior surface of the outer lip.

Inhabits Asia and Africa.

Fig. 5. M. *Javanus.*

Shell tapering, striate, with one row of oblique knots; anterior part of the outer lip with an inden-tation.

Lister Conch. Tab. 915. Fig. 8.

Specimen cinereous or pale livid, reddish towards the base, immaculate; beak rather long; suture of the whorls with a double tumid line; cleft in the lip deep.

Inhabits the Indian Ocean.

———

Fig. 6. M. *Vertagus*.

Shell having the upper whorls of the spire plaited; beak ascending, columella with one internal plait.

Lister Conch. Tab. 1020. Fig. 23.

Shell brownish white, aperture white, polished; whorls rounded, plaited on the upper margin, separated by a groove, lower ones nearly smooth; beak short.

Inhabits India.

———

PLATE XIX.

Fig. 1. TROCHUS *perspectivus*.

Shell convex, obtuse, margined; umbilicus pervious, crenulate.

Lister Conch. Tab. 636. Fig. 24.

Specimen white, with ochraceous spots on the margin and suture of the whorls; umbilicus funnel-shapel, surrounded by a spiral crenate margin;

shell flat underneath, longitudinally and transversely striate; acute margin of the whorls formed of a triple belt.

Inhabits the shores of Asia and Africa.

Fig. 2.   T. *Zizyphinus.*
Shell conical, smooth; whorls margined.
Lister Conch. Tab. 616. Fig. 1.
Specimen livid, with red indistinctly waved rays and spots upon the margin; transversely striate; whorls contiguous, upper ones finely granulate: aperture pearly.

Inhabits the European and African seas.

Fig. 3.   T. *dolabratus.*
Shell umbilicate, glabrous; columella recurved, twisted and plaited.
Lister Conch. Tab. 844. Fig. 72.
Shell white, with purplish brown transverse lines; whorls somewhat rounded, separated by a groove; columella with three folds.

Inhabits South America.

Fig. 4.   Turbo *muricatus.*
Shell umbilicate, subovate, acute, surrounded

with transverse striæ of raised dots; columellar margin obtuse.

Lister Conch. Tab. 30. Fig. 28.

Specimen purplish blue; aperture brown; whorls distant.

Inhabits Europe, South America, and Africa.

——

Fig. 5. T. *Chrysostomus.*

Shell subovate, rough, whorls surrounded with two rows of small arched spines.

D'Argenville Conch. Tab. 6. Fig. D.

Specimen yellowish and greenish white, irregularly rayed with brown; a row of angular brown spots between the rows of spines: columella and margin of the aperture white and pearly, within golden; whorls distant, transversely and longitudinally striate.

Inhabits India.

——

Fig. 6. T. *Anguis.*

Shell umbilicate, with transverse rounded striæ.

Martyn's Univers. Conch. 2. Tab. 70.

Specimen white, with dark green and blackish waved and angular marks; apex yellow; within pearlaceous; columella white; grooves obsolete on the under part; umbilicus circular, perforate.

Inhabits

—

Fig. 7. T. *scalaris*.

Shell conical, cancellate, or ribbed longitudinally; whorls distant.

D'Argenville Conch. Tab. 11. Fig. 5.

Specimen white, umbilicate, imperforate; no solid columella; whorls round, connected by the longitudinal, carinate, sub-oblique ribs, which are not continuous, but encompass each whorl separately, and adhere to each other at the upper and lower part; extremely thin and brittle.

Inhabits the coast of Barbary.

—

Fig. 8. T. *Terebra*.

Shell tapering; whorls with acute carinate striæ.

Lister Conch. Tab. 590. Fig. 4.

Specimen brownish white; whorls rather convex.

Inhabits European, African, and Chinese shores.

PLATE XX.

Fig. 1. HELIX *Scarabæus.*

Shell ovate, two-edged, sub-umbilicate; aperture toothed.

Lister Conch. Tab. 577. Fig. 31.

Specimen brown, variegated with pale spots, outer lip and teeth horny, white; whorls contiguous, double convex; aperture narrow, compressed, flexuous; each lip with three teeth.

Inhabits Asia.

———

Fig. 2. H. *cornea.*

Shell umbilicate, flat; whorls four, round.

Lister Conch. Tab. 197. Fig. 41.

Specimen blackish horn-colour, longitudinally striate, reverse; apex impressed.

Inhabits the fresh waters of Europe and Coromandel.

It is not easy to ascertain which is really the base, and which the apex of this shell; but if it be umbilicate, it must also be reverse: this appears upon the whole to be the fact, and the specimen is therefore drawn placed upon the perforated part;

and the necessary alteration is made in the terms of the specific description.

———

Fig. 3.   H. *nemoralis*.

Shell imperforate, rounded, smooth, diaphanous : aperture pyriform, or irregularly lunate.

Lister Conch. Tab. 57. Fig. 54.

Specimen pale orange, margin of the aperture brown ; whorls five.

Inhabits the woods of Europe.

———

Fig. 4.   H. *Columna*.

Shell tapering ; whorls contrary ; aperture oblong.

Lister Conch. Tab. 38. Fig. 37.

Specimen white, pellucid, with purplish longitudinal rays ; longitudinally and transversely striate.

Inhabits Guinea.

———

Fig. 5.   H. *stagnalis*.

Shell imperforate, ovate, subulate ; aperture ovate.

Lister Conch. Tab. 123. Fig. 21.

Specimen pale yellowish brown, darker towards

the apex; very finely striate longitudinally, very thin and brittle; body turgid; spire small; aperture wide; outer lip not margined.

Inhabits the stagnant waters of Europe.

Fig. 6. NERITA *glaucina*.

Shell smooth; spire rather obtuse; umbilicus half-closed; columellar lip gibbous, of two colours.

Lister Conch. Tab. 568. Fig. 19.

Specimen yellowish brown, with cinereous clouds; upper part of the whorls marked with short oblique reddish brown streaks; cinereous towards the apex; front of the columella, lip, and lower part of the aperture white, remainder ferruginous; umbilicus chesnut.

Inhabits Tranquebar and Barbary.

Fig. 7. N. *littoralis*.

Shell smooth; apex carious.

Lister Conch. Tab. 607. Fig. 39, 40.

Specimen yellow; very much resembling a Turbo in the form of the aperture, but a perfect Nerita in every other respect.

Inhabits the rocky shores of Europe.

Fig. 8. N. *striata*.

Shell thick, striate; spire rather prominent, columellar lip wrinkled, with three teeth, the upper one slightly lobed; margin and throat of the outer lip crenate; two teeth in the anterior part of the outer lip.

Martyn's Univers. Conch.

Specimen pale brown, yellowish green towards the apex; marked with shaded irregular rays of black; aperture white; throat yellowish; striæ numerous, rounded.

Inhabits

The author is not aware of this species having been described or named, although it be evidently figured in Martyn's Universal Conchology : he has therefore assumed its most remarkable character, though far from a peculiar one, as its nominal distinction.

———◆———

PLATE XXI.

Fig. 1. HALIOTIS *tuberculata*.

Shell sub-ovate; back transversely striate, rugged, and tuberculate.

Lister Conch. Tab. 611. Fig. 2.

M

Specimen whitish, mottled with olive green and black; inside pearlaceous; six orifices open.

Inhabits the European, Atlantic, and Indian Oceans.

The varieties of this species are very numerous, and differ much at the different stages of their growth. The specimen figured is less than usually tuberculate or plaited, and is somewhat flatter at the side; but still there is little doubt of its belonging to the species, as the striæ are precisely of the same description.

Fig. 2. (*interior and exterior*) H. *impertusa*.

Shell oblong, imperforate; very finely striate transversely and longitudinally; back convex.

Specimen rose-colour, clouded and speckled with cinereous, having four dusky bands interrupted by yellow angular spots and lines; inside pearlaceous.

This beautiful little shell is justly entitled to the designation *impertusa*, being not only imperforate, but having no impression of an orifice; yet will it hardly be denied a place in the genus Haliotis by any one who attentively inspects it.

Fig. 3. PATELLA *fornicata.*

Shell oval, recurved towards the apex; lip lateral, concave.

Lister Conch. Tab. 545. Fig. 33.

Specimen whitish, mottled and marked with brown waved lines, darker towards the tip and somewhat cinereous; lip white, prominent and lunate; margin acute.

Inhabits the Mediterranean and West Indies.

Fig. 4. *P. equestris.*

Shell orbicular, wrinkled outwardly; lip vertical, perpendicular.

Lister Conch. Tab. 546. Fig. 38.

Specimen white, hyaline, conical, obliquely truncate at the base, covered with irregular folds, wrinkles, and striæ; lip projecting, a little oblique, open on one side, situated under the centre of the vertex, formed like a canal.

Inhabits the Indian and American Seas.

Fig. 5. *P. granatina.*

Shell angular, broader at one end, with numerous muricate striæ.

Lister Conch. Tab. 534. Fig. 13.

Specimen outwardly white, with dark brown and chesnut spots and scales mostly angular; vertex brown; within like ivory, vertex with a large tawny brown spot square at one. end; margin brown; ribs many, unequal.

Inhabits Jamaica and Europe.

———

Fig. 6.   P. *Hungarica.*

Shell entire, conical, acuminate, striate; vertex hooked, revolute.

Lister Conch. Tab. 544. Fig. 32.

Specimen white, with a rosy tinge deeper towards the margin; finely striated longitudinally; transversely folded; base spreading.

Inhabits the Mediterranean, Adriatic, and American Seas.

———

Fig. 7.   P. *sanguinolenta.*

Shell ovate, convex, solid; striæ longitudinal, elevated, capillary, and flexuous; vertex surrounded with a broad punctate belt.

Martini Conch. 1. Tab. 7. Fig. 52.

Specimen white, with bright red striæ and dots,

variegated with scale-like spots of. brilliant white; yellowish towards the margin, vertex white, lateral.

Inhabits Africa.

Fig. 8. P. *Græca.*

Shell ovate, convex; margin crenulate within.

Lister Conch. Tab. 327. Fig. 1, 2.

Specimen white, with ten dull red rays; thick, covered with decussate striæ; internally white and smooth; perforation oblong.

Inhabits the shores of Europe.

PLATE XXII.

Fig. 1. DENTALIUM *elephantinum.*

Shell with ten angles, somewhat bent, striate.

Lister Conch. Tab. 547. Fig. 1.

Specimen pale green, obscurely annulate on the outside, with darker shades; apex white; within smooth; ribs projecting.

Inhabits the Indian and European seas.

Fig. 2. SERPULA *lumbricalis.*

Shell round, flexuous; apex spiral, acute.

Lister Conch. Tab. 548. Fig. 1.

Specimen white, brownish towards the tip; longitudinally wrinkled and transversely rugose.

Inhabits the Adriatic, Atlantic, and Indian Seas.

Fig. 3. S. *aquaria*.

Shell round, straight; circumference of one extremity radiate; disk furnished with cylindrical pores.

Lister Conch. Tab. 548. Fig. 3.

Specimen white, nearly smooth, gradually attenuated, smaller end open; larger one closed, convex.

Inhabits India.

Fig. 4. TEREDO *navalis*.

Shell thin, cylindrical, smooth.

Rumph. Mus. Tab. 41. Fig. F. G.

Specimen white, flexuous, rather tapering.

Fig. 5. SABELLA *Chrysodon*.

Shell subcylindrical, papyraceous, formed of testaceous fragments.

Martyn Conch. 1. Tab. 4. Fig. 29, 30.

Specimen dirty yellow; the particles of sand and shells adhere to a film, which is flexible when wet.

Inhabits Europe, India, and the Cape of Good
Hope.

The figure given in the plate is not an entire
specimen, but a part only, to show the general
appearance of the species.

———◆———

PLATE XXIII.

Fig. 1. BUCCINUM *Lepas.*

Shell oval; transversely ribbed, longitudinally
and transversely striate; ribs imbricate and tuberculate; spire small; aperture large, patulous.

D'Argenville Conch. Tab. 2. Fig. D.

Specimen brown, paler, and clouded with white
towards the apex, darker towards the margin; ribs
spotted with white; inside white, margin brown,
crenate with two teeth on the lower part; columellar lip flattened, projecting, somewhat revolute.

Inhabits Chili and the Falkland Islands.

This singular shell, which is denominated
PATELLA *Lepas* by Gmelin, has been selected as
a remarkable instance of doubtful generic character.
Conchologists are far from being agreed upon its
proper station. Without, therefore, presuming to

decide so nice a point for better judges than himself, the author has ventured to follow his own conviction, and to figure the specimen under the title of a Buccinum. His reasons for doing so have been,—that the canal is perfectly distinct, following the revolutions of the whorls,—and that this conformation is not to be found in spiral Patellæ, but precisely resembles that of Buccinum *patulum*, from which the present species appears to be removed but a single gradation in any one respect. At any rate it answers the intended object, which was to exemplify a *doubtful* shell, for the exercise of the learner.

Fig. 2. MUREX *dentatus*.

Shell obovate, caudate; whorls of the spire striate; apex produced very smooth; columella three-plaited.

Lister Conch. Tab. 815. Fig. 25.

Specimen white with bands of small reddish and ochraceous parallelogramic spots; whorls transversely undulate, crowned; apex cylindrical; three upper circumvolutions solid; aperture white, smooth.

Inhabits Tranquebar and Ceylon.

It is without any hesitation that this species is transferred from Voluta to Murex. The outline of the aperture is so absolutely characteristic of the latter, that it neutralises the claim of the toothed columella, which is not confined to the former. The teeth are nearly horizontal, and not oblique, as is usual in Voluta. No species can better illustrate the plaited section which has been proposed in the description of the genus Murex, than this, which possesses no property in common with its present congeners but its teeth. The specific name is changed from *Pyrum* to *dentatus*, because there is already a Murex *Pyrum*.

———

Fig. 3.　Helix *distorta*.

Shell solid, subumbilicate, striate, distorted, obtuse at the apex; body gibbous; aperture compressed, lunate.

Chemn. Conch. 5. Tab. 160. Fig. 1513. a. b.

Specimen white, polished; striæ oblique and curved; margin of the aperture thickened; last whorl produced into an obtuse projection on the right side.

Habitat unknown.

The general appearance of the shell bespeaks it a Helix, and not a Trochus as it has hitherto been designated. The convexity of the columella, and the upright situation and form of the aperture, irresistibly confirm the first impression.

———◆———

PLATE XXIV.

This plate is intended to point out to beginners in Conchology the great difference which exists, in certain species, between the shell in its young, unfinished state, and the same arrived at its perfection. The examples adduced are among the most conspicuous, but there are many others of which it would be very beneficial to students to acquire an early knowledge.

The mature specimen of Buccinum *Vibex* is given in Plate XXI. Fig. 3.

PLATE XXV.

Sections of the several genera are here offered rather as subjects of curiosity than as necessary to the science. They, however, show more accurately the true form of the inner chamber than even the aperture itself, which is often altered by the expansion or contraction of the margin, and is generally oblique to the direction of the whorls. The sections of Nautilus and Turbo constitute the two explanatory figures at the top of Plate I.

PLATE XXVI.

Fig. 1. HELIX *acutangula*.

Shell imperforate, thin, pellucid, acutely carinate; spire slightly convex; aperture ovato-lanceolate.

Nondescript.

Specimen pale horn-colour, finely striate longitudinally, whorls three, nearly flat at top; extremely brittle.

Habitat unknown.

This and the following shell, as well as the three

figured in Plate XXVII. and the last in Plate XXVIII., are all in the possession of Mrs. Mawe. They are certainly rare, and some of them are supposed unique: if either of them have been previously named or figured by any other writer, it is not within the knowledge of the author, who is far from wishing to lay claim to originality not his own, or to impose his names in preference to such as have been already conferred. In selecting the names he has endeavoured to find those which shall express some remarkable character of the species, and which may in some degree distinguish them from others by the meaning they convey, and not merely by their verbal sound.

Fig. 2. BUCCINUM *Pseudodon.*

Shell oblong, solid, transversely striate; whorls obtusely carinate, surrounded with transverse raised bands; aperture crenate, columella with two plaits.

Nondescript.

Specimen white, covered with a transparent brown epidermis; raised bands dark brown; aperture white, ovate; a few elevated striæ above the plaits on the columella, which is rather impressed towards the upper part; lower part of the outer lip furnished with the rudiments of a spine or tooth.

Habitat unknown.

The specific name Pseudodon has been assigned to this shell because the thickened internal rib, which is situated in the same part as the spine in B. Monodon, does not project beyond the margin, and has the appearance of a *false* rather than of a perfect *tooth*.

———

Fig. 3.   B. *strombiforme*.

Shell oblong, longitudinally plaited; outer lip separated above, expanded, thickened; aperture toothed.

Nondescript.

Specimen dark brown with a reddish tinge, aperture white; upper part of the columella and margin of the outer lip brown.

Shell smooth, plaits broad, folded towards the aperture; upper margin of the whorls surrounded with a tumid belt; lower part of the body transversely sulcated; columella 4-toothed, outer lip with nine teeth; apex pale, diaphanous.

Habitat unknown.

For permission to make a drawing of this and of Fig. 4, the author begs to acknowledge his obligation to Lady Wilson, of Charlton House.

—

Fig. 4. CHITON *spinosus.*

Shell 8-valved, semi-granulate : margin spinous. *Nondescript.*

Specimen fuscous, nearly black ; keel of the six middle valves marked with a cordate red spot; marginal membrane brown and broad; spines black, long, solid, subulate, and rather bent; lateral triangles on the valves granulate, remaining areas transversely striate ; posterior valve entirely covered with raised dots, internal surface whitish.

Habitat unknown.

This shell, which constitutes most unquestionably a distinct species of Chiton, was lately brought from Paris, but with it no information of the place from whence it originally came. If it were not unknown to naturalists, it is scarcely credible that so extraordinary a production should have been omitted in modern catalogues. It is a valuable acquisition to its genus, and is indeed well worthy of being placed at the head of an *armed division.*

PLATE XXVII.

Fig. 1. TURBO *madreporoides*.

This shell, if such it be, was lately sold with two others of the same description, at the sale of Lord Bute's collection, and was purchased by Mrs. Mawe. The species was brought originally, it is believed, from the coast of Africa, by Captain Young.

It is certainly extremely doubtful even to what Order this curious specimen belongs. Upon a minute examination of its external characters, and of such internal parts as have been perforated by marine insects, we need not hesitate to pronounce what it is *not;* what it really is, must still remain to be determined. The calcareous matter of which it is composed seems to be agglutinated in a different manner from that of Madrepores and Corals, to which it bears the most remarkable resemblance. The olive-green spots, which are often raised above the yellowish white surface, appear of the same construction with it, and are not regularly stellated, or more deeply pored. The arms are solid, and penetrated with the green colour in spots throughout. These processes vary in different specimens both in number and in length. The aperture is distinctly of the same shape as that of a Turbo, and hence the opinion may have arisen that the whole is parasitically formed by a coral insect which takes a certain species of Turbo for its nucleus. In opposition to this it may be stated, that there is no outward appearance of spire beyond the first whorl; that the aperture bears every mark of the recent passage of an inhabitant, being smooth, and free from all obstruction. That there is indeed no foreign shell is plain, from the circumstance of the substance being of the same nature in every part, and the inside of the outer lip showing the foundation of the spots through a thin slimy covering.

What then is it to be denominated? This will hardly be answered quite satisfactorily till the inhabitant be known, and till it be observed whether he have any organs with which to build this coral-like edifice. That it is however entirely the work of a molluscous worm, upon the whole appears most consonant with reason. If this be the case, it must, in the Linnæan system, be a Turbo, and it is proposed to call it specifically *madreporoides.*

At any rate it constitutes one more link in the great chain of nature, that which connects Testacea with Zoophyta.

––––

Fig. 2. TROCHUS *bifasciatus.*
Shell thin, pellucid, pyramidal, imperforate, whorls acutely carinate.
Nondescript.
Specimen white, with two purplish brown transverse bands and a dark brown apex.
Inhabits Pernambuco.

––––

Fig. 3. HELIX *gibberula.*
Shell conical, imperforate, gibbous; aperture toothed, compressed.
Nondescript.
Specimen reddish white, with a pale brown cloud and mark on the columellar side, and two brown bands upon the base; aperture white, with seven teeth, two nearest the exterior margin bent; outer lip margined, acute;—whorls six, smooth.
Inhabits Pernambuco.

PLATE XXVIII.

Fig. 1. CHITON *porosus.*

Shell 8-valved, carinate, valves with a tooth on each side, covered entirely with the marginal membrane.

Specimen pale olive above, sides brown; coriaceous covering cinereous or pale brown, tomentous; perforated over the back of each valve with a small slit and two tubular pores, one on each side; over the anterior valve four pores; valves finely striate and irregularly granulate; posterior valve indented in the margin; lateral triangles marked by an obsolete fold, and terminated by a small, sharp, tooth-like process.

Habitat uncertain, but probably New South Wales.

The animal possessing this very curious multivalve shell differs from the inhabitant of Chiton in the arrangement of the lungs, which do not extend so far on either side, but only about one third of the length; and in the intestinal canal. This dissimilarity in the worm will not however be a sufficient ground for separating it from the first

Linnæan genus. Nor will the extension of the membrane over the whole surface exclude it; for a regular gradation through the next species and C. *tunicatus* may be traced from this to any individuals of the genus which are perfectly uncovered.

The specimen figured in the plate and the following one are both deposited in the British Museum. They have been examined by Dr. Blainville of Paris, by whom a communication respecting them has, it is understood, been made to the French Philomatique Society. The names he has affixed to the two species are Cryptoconchus *porosus*, and C. *larvæformis*.

To the politeness of Dr. Leach, of the British Museum, the author owes his opportunity of making the annexed drawings and of inspecting the mollusca.

Fig. 2, 3, 4. C. *larvæformis*.

Shell 8-valved, valves without any keel, dissimilar, covered partially with the membrane.

Specimens having the three posterior valves white and smooth, bluish flesh-coloured at the sides, the others brown and striated; coriaceous membrane rough, cinereous olive, with a few scattered hairs. The disconnected valves are placed at the precise distance at which they were situated from each other when they were detached from the animal, which is preserved in spirits.

The dried specimen, which is figured in two positions, is probably much contracted, as three of the valves are imbricated : this shows, also, the portion of the shell which is exposed, about one third of each valve. The pores exist in this as well as the preceding species, but are much smaller, and scarcely discernible except when they are artificially distended. There is no visible communication between the external perforations and the lungs ; it is not easy therefore to determine their use or functions. The fact of the valves not being at all times approximate, will scarcely prevent this species from being justly ranked in the genus Chiton, as the valves are constructed so as to lie partially one over the other, and from the muscular contraction of the animal are doubtless often so arranged. But it can hardly be denominated Cryptoconchus, as a considerable part of the *shell* is not *hidden*. The two species *porosus* and *larvæformis* are however obviously of the same genus.

THE following List of English Trivial Names will be found useful to purchasers of shells, as dealers most frequently adopt them. It is, however, very much to be wished that all names should be abolished which are not strictly translations of the Linnæan specific ones. The trivial terms are mostly derived from some supposed resemblance to well-known objects ; and are therefore liable to be altered according to the fancy of collectors and catalogue-makers, who, in fact, have contrived to create as much uncertainty and confusion as could be expected from such a source. There are doubtless many technical distinctions which are not included in the subjoined enumeration, but the greater part of those in common acceptation are so. None of those are, of course, inserted which are merely rendered from the Latin, as "*scaly* Chiton" from Chiton *squamosus,* except in a few instances where the word is a substantive and not an adjective.

TRIVIAL NAMES.

——◆——

Linnæan Name.	English Name.
CHITON	COAT OF MAIL.
Minimus	Mealy C.
Asellus	Millepede.
LEPAS	ACORN SHELL.
Balanus	Common A. S.
Balanoides	Small, striated Do.
Diadema	Whale Do.
Tintinnabulum	Tulip Do.
Pollicipes	Cornucopia.
Anserifera	Striated A. S.
Anatifera	Barnacle or Goose shell.
PHOLAS	PIERCE STONE, OR PIDDOCK.
Dactylus	Long or prickly Piercer.
Costata	Large American Ph.
Striata	Goose-winged Ph.
Candida	Pur.
MYA	GAPER.
Prætenuis	Spoon-hinge.

Pictorum	Fresh-water Pearl Muscle.
Margaritifera	Pearl Gaper.
Perna	Smooth Muscle.
Syrmatophora	Angular G.
Glycymeris	Great M.

SOLEN	RAZOR OR SHEATH SHELL.
Siliqua	Long, brown S.
Ensis	Scimetar S.
Cultellus	Kidney Do.
Radiatus	Violet or radiated Do.
Strigilatus	Black Razor Do.
Castrensis	Zig-zag Do.
Pinna	Semi-oval Do.

TELLINA {	TELLEN OR DOUBLE WEDGE SHELL.
Gargadia	Toothed T.
Gari	Varying T.
Ferröensis	Carnation T.
Pectinata	Lister's T.
Remies	Waved T.
Virgata	Tulip Wedge.

| CARDIUM | HEART SHELL, COCKLE. |
| Medium | Marbled Heart. |

Costatum	White-fluted Heart.
Cardissa	Venus Heart.
Hemicardium	Ditto with smooth edge.
Aculeatum	Knotted H.; Ox H.
Echinatum	Rake H. Shell.
Isocardia	Rasp Cockle.
Edule	Common Do.
Fragum	White Strawberry Do.
Æolicum	Janus.
Unedo	Strawberry Do.
Magnum	Yellow, ribbed Do.
Retusum	Diana Heart.
Rusticum	Bear's Paw; Tufted H.

MACTRA	MACTRA.
Lutraria	Large M.

DONAX	WEDGE SHELL.
Scortum	Beaked W.
Trunculus	Yellow Do.
Faba	Bean-shaped Do.
Irus	Foliated Do.

VENUS	VENUS SHELL.
Dione	Prickly-mouthed V.
Paphia	Oldwoman; wrinkled V.

Fimbriata	Cancellated or Chequered V.
Histrio	Map V.
Puerpera	Spotted V.
Deflorata	Purple Streak V.
Pectunculus	Painted V.
Casina	Broad-ribbed V.
Exoleta	Painted V. or Cockle.
Literata	Camp-lettered V.
Circinata	Compass V.
Erycina	Polished V.
Chione	Smooth Brown V.

SPONDYLUS   SPONDYLE.

Gædaropus	Red thorny Oyster; Ass's Hoof.

CHAMA   CHAME.

Cor	Fool's Cap C. or Cockle.
Gigas	Furbelow'd Clamp.
Hippopus	{ Cabbage-leaf Chame, or Bear's Paw Clamp.
Arcinella	Hedge Hog.

ARCA   ARK SHELL.

Noæ	Noah's Ark.
Cucullus	Chambered A.
Antiquata	Jamaica A.

Undata	Lettered Ark.
Glycymeris	Orbicular Do.
Pectunculus	Spotted Do.
Nucleus	Silvery Do.

OSTREA	OYSTER.
Maxima	Common English Scallop.
Jacobæa	Mediterranean Do.
Radula	Ducal Mantle; Rasp.
Pallium	Royal Mantle.
Nodosa	Duck's-foot.
Pleuronectes	Compass Scallop; Sole.
Pusio	Wrinkled S.
Lima	The File.
Opercularis	Painted S.
Malleus	Hammer Oyster.
Folium	Leaf O.
Edulis	Common eatable O.
Perna	Oblong O.
Isogonum	Rudder.
Ephippium	Saddle-oyster.
Vulsella	Tongue-shaped O.

ANOMIA	ANOMIA.
Ephippium	Green onion rind A.

Cepa	Onion.
Electrica	Small Amber A.
Placenta	Chinese Window Oyster.
MYTILUS	SEA MUSCLE.
Crista Galli	{ Tree Muscle. Hog's-ear. Cock's-comb.
Hyotis	Great Finger M. Horned M.
Margaritiferus	Pearl Muscle. Mother-of-pearl.
Lithophagus	Brown M. or burrowing M.
Aristatus	'Cross-beak M.
Ungulatus	* Great striated Magellanic M.
Bidens	Furrow-cap M.
Modiolus	Smooth or Great M.
Hirundo	Swallow M.
Faba	Bean.
Morio	Mulberry.
PINNA	NACRE.
Muricata	Small red aculeated N.
Rotundata	Giant N.
Digitiformis	Small white N.
ARGONAUTA	{ PAPER SAILOR or PAPER NAUTILUS.
Argo	Oriental P. S.
Cymbium	Minute Do.

NAUTILUS	SAILOR.
Pompilius	Great chambered S.
Spirula	Ram's-horn S.
Lituus	Crozier.

CONUS	CONE SHELL.
Marmoreus	Black Tiger C.
Imperialis	Imperial Crown C.
Eburneus	Square Spotted C.
Generalis	Flambeau C.
Miles	Girdle; Bastard Admiral C.
Princeps	Persian Robe C.
Papilio β	Butterfly's Wing C.
Figulinus	{ Brown striped C. or Beech-wood C.
Ebræus	Rustic Music; Hebrew C.
StercusMuscarum	Flea-bitten C.
Monachus	Agate C.
Genuanus	Butterfly's Wing C.
Achatinus	Tulip C.
Textile	Gold Brocade C.
Luzonicus	Spotted Velvet C.
Vittatus	Ribbon C.
Classiarius	Sailor C.
Mercator	Net-work C.

Affinis	Commandant C.
Rosaceus	Aurora C.
Aulicus	Porphyry C. or Brunette C.
Nussatella	Shagreen C.
Spectrum	Spectre C.
Geographicus	Silk Brocade C.

CYPRÆA   GOWRY or COWRY.

Exanthema	False Argus.
Arabica	Nutmeg Cowry.
Argus	The true Argus.
Carneola	Burnt mouth'd C.
Talpa	Mole or burnt mouth C.
Lurida	Mouse.
Tigris	Leopard C.
Scurra	Green spot C.
Caput Serpentis	Viper's Head.
Mauritiana	Lesser Surinam Toad.
Vitellus	Fallow Deer.
Isabella	Orange-tipt C.; Porcelain.
Asellus	Wasp.
Moneta	Trussed Fowl; Common Money C.; Blackmoor's-teeth.
Caurica	Dark spotted C.
Erosa	White spotted C.

Pediculus	Sea Louse.
Nucleus	Wrinkled C.
Staphylæa	Wood Louse.
Dracæna	Dragon C.
Helvola	Star C.
Globulus	Pearl C.

BULLA · DIPPER.

Ovum	Poached Egg.
Volva	Weaver's Shuttle.
Birostris	Bastard Weaver's Shuttle.
Spelta	Oblong D.
Gibbosa	{ Short Gibbous Shuttle ; Gondola.
Naucum	Sea Nut.
Ampulla	Pewit's Egg.
Lignaria	{ Open-mouthed Nut or Wood-Dipper.
Hydatis	Paper D.
Amygdalus	Almond.
Akera	Elastic D.
Physis	Striped D.
Amplustre	Banded D.
Ficus	Fig.
Terebellum	Auger.

Fontinalis	Fresh-water D.
Virginea	Orange Flag. .
Zebra	Zebra Chersina.
Achatina	{ Broad-striped Zebra, or Pink-mouthed Chersina.
VOLUTA	RHOMB SHELL or CYLINDER.
Auris Midæ	Midas' Ear.
Porphyria	{ Large clouded Rhomb or Camp Olive.
Oliva	Yellow Rhomb; Olive.
Gibbosa	Clouded Olive.
Ispidula	Enamelled O.
Pinguis	Quaker O.
Utriculus	Gibbous O.
Dactylus	Six-plaited O.
Persicula	Red spotted O.
Paupercula	Zebra Rhomb.
Mitra Episcopalis	Bishop's Mitre.
Papalis	Papal Mitre.
Musica	West India Music Shell.
Vespertilio	Wild Music; Bat.
Ebræa	Oriental Music.
Rustica	Net Olive.
Capitellum	White Music.

Ceramica	Larger Devil.
Scapha	Lightning.
Corona	Ducal Crown.
Pyrum	Turnip.
Æthiopica	{ White-mouth'd Melon; Æthiopic Crown.
Cymbium	Clouded or Boat Melon.
Olla	Melôn.
Navicula	Gondola.
BUCCINUM	WHELK.
Galea	Brown Tun.
Olearium	Tun.
Perdix	Partridge Tun.
Dolium	Spotted Tun.
Cornutum	{ Thimbled Helmet; Triangular Whelk.
Rufum	Red or Bull's mouth Helmet.
Tuberosum	Casket.
Flammeum	Triangular Casket.
Testiculus	Bonnet Casket.
Areola	Small Dice C.
Saburon	Gray C.
Vibex	Agate C.
Arcularia	Fingers and Thumbs.

Glaucum	Bezoar Helmet.
Rugosum	Zebra Helmet.
Harpa	Musical Harp.
Monodon	Unicorn.
Persicum	Pers. Music or Necklace Scoop.
Patulum	Mulberry; Scoop.
Spiratum	Joppa Whelk.
Lapillus	Purple Straining W.
Nucleus	Small W.
Piscatorium	Knobbed W.
Maculatum	{ The brown Mitre, the Marline-spike.
Subulatum	The Tiger Spire.
Duplicatum	Press Screw.
STROMBUS	SCREW.
Fusus	Spingle.
Pes Pelecani	Pelican's Foot.
Chiragra	Devil's Claw.
Scorpius	Scorpion.
Lambis	Spider.
Millepeda	Millepede.
Lentiginosus	African pink Conch.
Auris Dianæ	Plough Frog.
Gallus	Plough.

Gibberulus   Pouter; Spotted Pouter.
Accipiter   Spinous S.
Epidromis   Mottled Fawn.
Canarium   Partridge.
Urceus   Pitcher.
Lucifer   Spiked Whelk.
Palustris   Marsh Club.
Gigas   Large Conch; large Roller.

   MUREX   CALTROP or ROCK SHELL.

Haustellum   Snipe.
Tribulus   { Thorny Woodcock, or Venus'
                      Comb.
Cornutus   Thorny Snipe.
Brandaris   Thorny Snipe's Head.
Trunculus   Tyrian Dye.
Lingua   Sheep's Tongue.
Ramosus   Purpura, Devil.
Femorale   Triangular Whelk.
Scorpio   Skeleton.
Saxatilis   Endive Shell.
Erinaceus   Urchin.
Rana   Thorny Frog.
Olearium   Oil Jar.
Lampas   Swiss Trowsers.

Ricinus	Spur Shell.
Nodus	Chesnut.
Fucus	Old Maid.
Neritoideus	Mulberry.
Mancinella	Ditto.
Colus	Spindle.
Canaliculatus	Bottle Whelk.
Tritonis	Trumpet.
Rubecula	Footman.
Melongena	Open-mouthed M.
Anus	Grimace.
Babylonius	Tower of Babel.
Perversus	Left-handed Fig.
Trapezium	Persian Robe.
Morio	Helmet.
Pusio	Wreath.
Amplustre	American Flag.
Aluco	Caterpillar; Hercules' Club.
Granulatus	Silk Worm.
TROCHUS	TOP SHELL; BUTTON SHELL.
Niloticus	Large-marbled Trochus.
Alveare	Bee-hive.
Conspersus	Poppy.
Fanulum	Pagoda.

Conchyliophorus   Carrier.
Pumilio   Dwarf.
Lubeo   Double-lipped T.
Argyrostomus   Ink Horn.
Ziziphinus   Livid T.
Granatum   Tiger.
Mauritianus   Great toothed T.
Dolobratus   Zebra.
Perspectivus   Staircase.
Solaris   Sulcated. Sun.
Pharaonius   Strawberry T.
Telescopium   Telescope.

TURBO   WHORL or WREATH.

Petholatus   { High-headed Oriental Ribband
                     Snail.
Pagodus   Pagoda.
Littoreus   Periwinkle.
Pullus   Painted W.
Coclus   Spotted Silver Mouth.
Chrysostomus   Golden-mouthed Snail.
Cidaris   Turban.
Calcar   Spur.
Rugosus   Large Silver-mouthed W.
Setosus   Leopard.

Anguis	Snake.
Marmoratus	Large Green W.
Pica	Magpie.
Delphinus	Dolphin.
Mumia	Double-toothed W.
Scalaris	Wentletrap.
Terebra	Tambour Needle.
Turritella	Spindle.
Exoletus	Ribbed Screw.
Clathrus	Bastard Wentletrap.
Acutangulus	Press Screw.

HELIX	SNAIL.
Scarabæus	Witch or Cockchafer.
Haliotoidea	White-ear Snail; Venus' Ear.
Lapicida	Rock S.
Ringens	Tooth'd Lamp.
Avellana	Hazel Nut.
Cornu Arietis	Ram's Horn.
Urceus	Cocoa Nut.
Ianthina	Violet S.
Stagnorum	Barley Corn.
Ampullacea	Smooth-girdled S.
Pomatia	Vineyard Snail; Edible S.
Zonaria	Ribband Do.

NERITA	NERITE.
Mammilla	White-nipple Nerite.
Histrio	Guinea-hen N.
Chamæleon	Hoofs.
Exuvia	Deep-ridged N.
Pulligera	Red Nerite.
Stella	Black-rayed N.

HALIOTIS	SEA EAR.
Tuberculata	Common Ear Shell.
Asinina	Ass's Ear S.
Midiæ	Leafy Ear S.

PATELLA	LIMPET.
Equestris	Cup and Saucer L.
Fornicata	Slipper L.
Neritoidea	Chambered L.
Chinensis	Chinese Bonnet L.
Saccharina	Star L. Astrolepas.
Goreensis	Sandal.
Crepidula	Transparent L.
Repanda	Small Sum.
Margaritacea	Great Sun.
Barbara	Toothed L.
Oculus	Goat's-eye.

Granularis	Striated L. Thorny L.
Granatina	Garnet L. or Carbuncle.
Testudinaria	Buckler; Tortoise L.
Rustica ·	Dutch Bonnet; Boot.
Pustula	Mask L.
Chlorosticta	Pigeon's Throat.
Ulyssiponensis	Buckler.
Calyptra	Helmet.
Cassida	Lentil Seed.
Pellucida	Blue-rayed L.
Ferruginea	Bronze L.
Sanguinolenta	Beauty L.
Compressa	White L.
Laciniosa	Double-eyed L.
Lusitanica	Auricula.
Fissura	Cracked L.
Mitrula	Cap. L.
Umbellata	Chinese Parasol.
Pustula	Doubtful L.
Pileolus	Open Cap.
Hungarica	Fool's-cap L.
Mammillaris	Black Hair Streak L.
Cypria	Mushroom L.

DENTALIUM   TOOTH SHELL.

Elephantinum { Fluted Elephant's Tooth ; Horn-
                green Pencil.

Entalis   { Striated Tooth-shell, or Dog's
                Tooth-shell.

Gadus   Hake's T. S.

Trachea   Windpipe.

SERPULA   WORM SHELL.

• Lumbricalis   Spirally twisted.

Stellaris   Rayed Pin's Head.

Anguina   Serpent.

Semilunum   Small Seed.

Aquaria   Watering-pot.

Arenaria   Oven Shell.

Cereolus   Bougie.

Volvox   Caterpillar.

Ocrea   Boot.

THE author begs to preface the following Catalogue of Testaceological Writers, by an acknowledgment of his being indebted for the substance
of it to a paper in vol. vii. of the Linnæan Trans-
actions, by Dr. Maton and the Rev. Mr. Rackett.
It is so perfect a compendium of the subject, that
he trusts he need offer no apology for wishing to
render it more generally useful, by putting it into
a form which is capable of wider dissemination
than the work in which it is at present extant. In
altering the arrangement from a chronological into
an alphabetical one, he by no means wishes to
imply that he considers the former method excep-
tionable, but only less adapted to an elementary
treatise, intended rather for the horn-book of
learners, than for the information of the scientific.

THE numerals which are placed after the name
of the conchological author refer to the nature of
the work, the title of which is given immediately
afterwards in Italics. By these numerals will be
designated the branch of science which each writer
has pursued; and by comparing them with the
annexed arrangement, it will be known at once

what sort of instruction is to be expected from his labours.

Testaceological writers are to be classed as follows :

I. Those who have written generally of Conchology.

II. Those who have described a single genus, family, or species.

III. Those who have described the shells of certain parts of the globe.

IV. Those who have described museums or collections of shells.

V. Those who have described minute shells, microscopic subjects.

VI. Those who have detailed the wonders of the science.

VII. Those who have written on the anatomy of testaceous worms.

VIII. Those who have written on their physiology.

IX. Those who have invented or followed a systematical arrangement of shells.

X. Those who have commented on the works of others.

XI. Those who have published plates illustrating Conchology.

WRITERS ON CONCHOLOGY.

A

ADAMS JOHN. III. V. *Description of Minute Shells found on the Coast of Pembrokeshire.* "Trans. Linn. Soc." vol. iii. p. 64. 68. 252. 254. vol. v. p. 1—6.

ADANSON MICH. II. III. IX. *Déscription d'une nouvelle Espèce de Ver qui ronge Bois et les Vaisseaux, observée en Senegal.* "Mém. de l'Acad. des. Sc." 1759. p. 249-278, with plates.

Histoire Naturelle du Senegal. Paris, 1757. 4to. with 19 plates.

ÆLIAN. I. Died about A. D. 140.

Πιρὶ Ζώων Ἰδιότητος.

ALBERTUS MAGNUS. I. *De Animalibus.* Venet. 1795. fol.

ALDROVANDUS. I. *De Mollibus, Crustaceis, Testaceis, et Zoophytis.* Vol. iii. Bononiæ, 1606. fol. with wooden cuts.

ARISTOTLE. I. Died about 322. A. C.

Πιρὶ Ζώων Ἰστορίας τὸ Δ. κιφ. ἰ.

The History of Animals.

B.

BARRELIER JACOBUS. III. *Specimen Insectorum quorundam Marinorum,* in Libro de Plantis per Galliam, Hispaniam, et Italiam observatis. (A. de Jussieu.) Paris, 1714. fol. with plates.

BARTRAM JOHN. III. *Observations concerning the Salt-marsh Muscle, the Oyster-banks, and the Fresh-water Muscle of Pennsylvania,* in " Phil. Trans." vol. xliii. p. 157-159. (1744.) with figures.

BASTER JOB. VIII. *Opuscula Subseciva.* Haarl. Lib. i. 1759. 4to. with copper plates.

BELON PIERRE. I. *De Aquatilibus.* Lib. ii. Paris, 1553. 8vo. p. 448. with plates.

BESLER MICH. RUPERT. IV. *Gazophylacium Rerum Naturalium.* 1642. Lips. 1733. fol. with plates.

BESLER BASIL. XII. *Fasciculus rariorum et Aspectu digniorum varii Generis,* 1616. fol.

BOCCONE PAOLO. VIII. *Observazioni Naturali.* Bologna, 1684. 12mo.

BONNET CHARLES. VIII. *Exp. sur la Régénération de la Tête du Limaçon terrestre,* in the " Journ. de Physique," tom. x. p. 165—179. (1775.)

BONVIÇINI GIUSEPPE. VIII. *Lettera al Sign. Prof. Girardi*, in " Mem. della Soc. Ital." tom. vii. p. 291—299. (1794.)

BORLASE WILLIAM. III. *Natural History of Cornwall.* Oxford, 1758. fol. with plates.

BORN IGNAZ EDLER VON. IV. *Index Rerum Naturalium Musei Cæsarei Vindobonensis. Pars. 1ma. Testacea.* (Latine et Germanice.) Vindobonæ, 1778. 8vo.

   Testacea Musei Cæsarei Vindobonensis. Vindob. 1780. fol. with finely coloured plates.

BOSC. L. A. G. IX. *Histoire Naturelle des Coquilles*, 5 vols. Paris, 1802. 12mo. with plates.

BOYLE ROBERT. VIII. *Of some Phænomena afforded by Shell-fishes*, in " Phil. Trans." vol. v. p. 2023. (1670.)

BOYS WILLIAM. III. V. *Testacea minuta rariora nuperrimè detecta in Arenâ Littoris Sandvicensis.* Lond. 1784. 4to. with 3 plates.

BRACHIUS JACOBUS. VIII. *De Ovis Ostreorum*, in " Eph. Ac. Nat. Cur." dec. 2. an. 8. p. 506. (1690.)

BRADLEY RICHARD. I. *Philosophical Account of the Works of Nature.* London, 1721. 4to. with figures.

BREYNIUS JOH. PHIL. I. III. IX. *De quibusdam Conchis minùs notis*, in " Mem. sopra la

Fisica e Istoria Naturale," tom. i. p. 175.
(Lucca, 1743.)

Epistola varias Observationes continens in Itinere per Italiam suscepto anno 1703, in "Phil. Trans." vol. xxvii. p. 447—459.

Dissertatio Physica de Polythalamiis. Gedan. 1732. 4to. p. 64. with plates.

BRISSON. II. *Observations sur une Espèce de Limaçon terrestre dont le Sommet de la Coquille se trouve cassé sans que l'Animal en souffre,* in the "Mém. de l'Acad. des. Sc." 1759. p. 99—114. with 13 figures.

BROCCHI G. III. *Conchiologia fossile Subapennina.* Milan. 1814. 2 vols. 4to. with 16 plates.

BROOKES SAMUEL. IX. *Introduction to the Study of Conchology.* Lond. 1815. 4to. with coloured plates.

BROWN THOMAS. IX. *Elements of Conchology.* Lond. 1816. 8vo. with plates.

BBUCKMANN FR. ERN. II. *De curiosissimis duabus Conchis Marinis.* Brunsv. 1722. 4to. with a copperplate.

BRUGUIERE J. G. I. II. VIII. IX. *Histoire Naturelle des Vers,* tom. i. in "l'Encyclopédie Méthodique." (Paris, 1789—1792.) Livraison 32 et 48.

In the "Journ. d'Hist. Nat." tom. i.

p. 20. (Paris, 1792. 8vo.) tom. i. p. 103. 339.

Sur la Formation de la Coquille des Por-cellaines, et sur la Faculté qu'ont leurs Ani-maux de s'en détacher, et de les quitter à des différentes Epoques, in the " Journ. d'Hist. Nat." tom. i. p. 307 and 321.

BRUNNICH MARTIN THRANE. IX. *Fundamenta Zoologica.* Hafn. et. Lips. 1772. 8vo. p. 253.

BUONANNI FILIPPO. I. IV. *Ricreazione dell' Occhio e della Mente nell' Observationi delle Chioc-ciole.* Rom. 1681. 4to. with figures.

Recreatio Mentis et Oculi in Observatione Animalium Testaceorum. Rome, 1684. 4to. with copperplates.

Supplementum Recreationis, &c. in parte 2da.

Observationum circa Viventia quæ in Re-bus non viventibus reperiuntur. Rome, 1691. 4to. with 10 copperplates, not before pub-lished.

Musæum Kircherianum. Rome, 1709. fol. with copperplates.

BYTEMEISTER HEN. JOH. XII. *Bibliothecæ Appen-dix.* Ed. 2da. Jul. 1735. 4to. with copper-plates.

C

CHARLTON WALTER. X. *Onomasticon Zoicum.* Lond.
1668. 4to.

CHEMNITZ J. HIERON. I. II. *Neues Systematisches
Conchylien Cabinet*, from vol. iv. 1780, to
vol. xi. 1795. Nurenberg; with the Index
or Register to the first 10 volumes. Nurenberg, 1788.

　*Observationes de Testaceis multivalvibus
nonnullis*, in " Nov. Act. Nat. Cur." t. viii.
Ap. 35—42. (1791.)

　De Chitonibus. Nurenberg. 1788, with
plates.

CHICCO. IV. *Museum Calceolarium.* 1622. fol.

COHAUSEN JO. HEN. IX. *Conspectus Sciographicus
Testaceorum*, in " Dissert. Epistolicarum,"
tom. iii. (Francof. 1754.) 8vo. p. 296—346.

COLE WILLIAM. II. *Observations on the Purple
Fish*, in " Phil. Trans." vol. xv. p. 1278—
1286, with a plate, (1685.) *Purpura An-
glicana*, London, 1689. 4to.

COLUMNA FABIUS. I. II. *Aquatilium et Terrestrium
aliquot Animalium aliarumque Naturalium
Rerum Observationes.* Rome, 1616. 4to. with
copperplates, with notes by D. Major, M.D.
Killiæ, 1675. 4to. with wooden cuts.

Purpura, &c, Rome, 1616. 4to. with 7 copperplates.

CORDINER CHARLES. III. *Remarkable Ruins, &c. in North Britain.* London, 1788 — 1795. 4to. with coloured plates.

COTTE. VIII. In the " Journ. des Sçavans," 1770. et " Journ. de Physique," tom. iii. p. 370.

CUNINGHAME JAMES. III. *A Catalogue of Shells, &c. gathered at the Island Ascension,* in " Phil. Trans." vol. xxi. p. 295—298. (1699.)

CUVIER GEORGE. VII. *Anatomie de la Patelle commune,* in " Journ. d'Hist. Nat. tom. ii. p. 81—95. with one plate. (1792.)

CYPRIANUS JOHANNES. I. *Franzii Historia Animalium sacra,* cap. 8. Françof. et Lips. 1712. 4to.

D

DA COSTA EMANUEL MENDEZ. I. III. IX. *Conchology, or Natural History of Shells.* London, fol. 12 plates. (English and French.)

British Conchology. (Fr. and Eng.) London, 1778. 4to. with coloured plates.

Elements of Conchology. London, 1776. 8vo. with seven plates.

DALE. III. *History and Antiquities of Harwich,* by SILAS TAYLOR. 2d edit. Lond. 1732. 4to.

D'ARGENVILLE ANT. JOS. DESALLIER. I. IX. *L'Histoire Naturelle éclaircie dans deux de ces principales Parties, la Lithologie et la Conchyliologie.* Paris, 1742 et 1757. 4to. 1780. 2 vols. with plates.

D'AVILA. IV. *Catalogue Systématique et Raisonné,* vol. i. (1767, 8vo.) with 22 plates.

DE BERGEN CAR. AUG. IX. *Classes Conchyliorum.* Nuremb. 1760. 4to. pp. 132.

DE BONDAROY AUGUSTE DENIS FOUGEROUX. II. *Mémoire sur le Coquillage appellé Datte en Provence,* in the " Mém. Etrang. de l'Ac. Roy. des Sc." tom. v. p. 467—478. (1768.) with one plate.

DE HEIDE ANTON. VII. *Descriptio anatomica Mytili,* in " Act. Erud. Lips." 1684. p. 426.

DE JOUBERT. II. *Mémoire sur une Coquille de l'Espèce des Poulettes péchée dans la Méditerranée,* in the " Mém. Etrang. de l'Ac. des Sc." (1774.) tom. vi. p. 77—80, et p. 83—91. with figures.

DE REAUMUR RENE ANTOINE F. VIII. *De la Formation et de l'Accroissement des Coquilles des Animaux tant terrestres qu'aquatiques, soit de Mer soit de Rivière,* in the " Mém. de l'Acad." 1709. p. 364—400. with two plates.

De Rochfort. III. *Histoire Naturelle et Morale des Isles Antilles*, ch. 19. 2d edit. Rotterd. 1665. 4to. with plates.

De Ribaucourt M. VIII. *Sur la Génération des Buccins d'Eau douce*, in the " Journ. d'Hist. Nat." tom. i. p. 428.

De la Faille Clemens. II. *Sur l'Origine des Macreuses*, in the " Mém. Etrang. de l'Acad. Roy. des Sc." (1780.) tom. ix. p. 331—344. with one figure.

Des Hayes Lefebure. II. *Notices sur le Bœuf marin, autrement nommé Béte à huit Ecailles, ou octovalve*, in the " Journ. de Phys." 1787. tom. xxx. p. 209—214. with figures.

Delandes. II. *Eclaircissemens sur les Oiseaux de Mer*, in his " Recueil de différens Traitez de Physique et d'Histoire Naturelle," tom. i. p. 197. (Paris, 1736. 3 vols. 12mo.)

Sur les Vers qui rongent le Bois de Vaisseaux, in the " Recueil," &c. tom. i. p. 214.

Dicquemare. II. *Insectes marins Déstructeurs des Pierres*, in the " Journ. de Phys." tom. xviii. p. 222—224. tom. xx. p. 228—230.

Dillwyn L. W. IX. *Descriptive Catalogue of recent Shells*. London, 1817. 2 vols. 8vo.

Donovan Edward. III. *Natural History of British*

Shells. Lond. 1799. 8vo. five vols. with coloured plates.

DRAPARNAUD J. P. R. III. *Histoire des Mollusques de la France.* Paris, 1805. 4to.

DUCHÉSNE. XII. *Recueil des Coquilles fluviatiles et terrestres qui se trouvent aux Environs de Paris,* three plates, fol.

DUFRESNE. II. *Notice sur les Balanus,* in the " Ann. du Mus. Nat." Cahier 6. p. 465. with fig.

DU HAMEL HENRI LOUIS. II. *Quelques Expériences sur la Liqueur colorante que fournit la Pourpre, &c.* in " Mém. de l'Acad. Fran." 1736. p. 49—63.

DU MOLINET CLAUDE. IV. *Le Cabinet de la Bibliothèque de Ste. Geneviève.* Paris, 1692. fol. with copperplates.

DU PATY MERCIER. II. *Sur les Bouchots à Moules,* in the " Recueil de l'Acad. de Rochelle," 1752. p. 79—95. with three plates.

DU TERTRE JEAN BAPT. III. *Histoire Générale des Antilles habitées par les François,* tom. ii. (Paris, 1667. 4to.)

E

EDWARDS GEORGE. XII. *Gleanings of Natural History.* (English and French.) Lond. part 1. 1758. part 2. 1760. 4to.

ELLIS JOHN. II. *An Account of several rare Species of Barnacles*, in " Phil. Trans." vol. l. p. 845—855. (1758.) with figures.

F

FABRICIUS OTHO. III. *Fauna Groenlandica.* Hafn. et Lips. 1780. 8vo. with one plate.

FAVANNE DE MM. Pere et Fils. I. *La Conchyliologie de M. D'Argenville augmenté de planches.* Paris, 1780. 3 vols. 4to.

FAVART D'HERBIGNY. I. *Dictionnaire d'Histoire naturelle, qui concerne les Testacées, ou les Coquillages de Mer, de Terre, et d'Eau-douce.* (1775.) 3 vols. 12mo.

FEHR JOANNES MICH. II. *De Carina Nautili elegantissima*, in " Eph. Ac. Nat. Cur." Dec. 2. An. 4. p. 210. (1686.) with copperplates.

FICHTEL LEOPOLD. A. V. *Testacea Microscopica aliaque minuta ex Generibus Argonauta et Nautilus ad Naturam picta et descripta.* Wien. 1798. 4to. 24 copperplates coloured. (Latin and German.)

FISCHER CHRISTIAN GAB. IX. *Specialis Tabula Synoptica sistens Cochlides et Conchas*, in Kleinii " Disp. Echinodermatum," p. 73— 75. (Gedani, 1734. 4to.) and in the same

work by N. G. Leske, p. 60—62. (Lips.
1778. 4to.)

FORBES GEORGE. II. *A Letter relating to the Patella,
or Limpet Fish, found at Bermuda*, in " Phil.
Trans." vol. l. p. 859—860. (1759.) pl. 35.

FORSKAHL PETER. III. *Descriptiones Animalium
quæ in Itinere Orientali observavit P. F.*,
edit. a C. NIEBUHR. Havniæ, 1775. 4to.

FORTIS ALBERTO. III. *Viaggio in Dalmazia.* Venezia, 1774. 4to. with fig.

FULBERTI CONTI GODEFRIDO. VIII. *Riflessioni, &c.*
Roma, 1683. Bologna, 1695. 12mo.

G

GEOFFROY. III. *Traité sommaire dés Coquilles tant
fluviatiles que terrestres qui se trouvent aux
Environs de Paris.* Paris, 1767. 8vo. with 3
plates.

GERSAINT EDM. FRANÇOIS. IV. *Catalogue raisonné
dés Coquilles, &c.* Paris, 1736. 8vo.

GESNER CONRAD. I. X. XII. *Lib. 4. de Piscium et
Aquatilium Animantium Naturâ.* Tiguri,
1558. fol. Francof. 1620. with wooden cuts.

    Nomenclator Aquatilium Animantium. Ti-
guri, 1560. fol.

    Icones Animalium Aquatilium. Tiguri, fol.
1560.

GEVE NICHOLAS G. I. *Monatliche belustigungen
im reiche der Natur, au Conchylien und See-
gewachsen.* (Germ. et Gall.) Hamb. 1755.
4to. with 33 plates.

GINANNI CONTE GIUSEPPE. III. In his " Opere
Postume," tom. ii. (Venezia, 1757. fol.) with
38 plates.

GINANNI CONTE FRANCESCO, IV. *Produzioni Naturali che se ritrovano nel Mus. Ginanni in
Ravenna.* Lucca, 1762. 4to. with fig.

GOTTWALD CHRISTOPHER. XII. *Museum Gott-
waldianum,* Gedan. 1714. fol. with copperplates.

*Musei Gottwaldiani Testaceorum, Stella-
rum Marinarum, et Coralliorum quæ super-
sunt Tabulæ.* Nuremb. 1782. fol. with 49
copperplates. edit. by JOH. SAM. SCHROTER.

GREW NEHEMIAH. IV. *Museum Regalis Societatis,*
or Catalogue and Description of the Natural
and Artificial Rarities belonging to the Royal
Society, and preserved in Gresham College,
London, 1681. fol. with plates.

GRONOVIUS LAUR. THEODORUS. IV. XI. *Zoophy-
lacium Gronovianum.* Lugd. Bat. 1781. fol.
with copperplates.

GUALTIERI NICOLAI. IV. *Index Testarum Conchy-*

liorum quæ adservantur in Museo NICOLAI
GUALTIERI, *Philosophi et Medici, Floren-*
tini, &c. Florentinæ, 1742. fol. with copperplates.

GUETTARD JEAN ETIENNE. III. IX. *Sur le Sable*
Coquillier de Zalbach, in his " Mém. sur
différentes Parties des Sc. et Arts," tom. ii.
p. 21—22. (1770.)

Observations qui peuvent servir à former
quelques Caractères de Coquillages, in the
" Mém. de l'Acad." 1756. p. 145—183.

Sur le Rapport qu'il y a entre les Coraux
et les Tuyaux marins, et entre ceux-ci et les
Coquilles. Ibid. 1760. p. 114—146. with
plates. *Sur les Tuyaux marins*, in his
" Mém. sur différentes Parties d'Hist. Nat."
tom. iii. p. 18—208. (1770.)

H

HANNEMAN JOH. LUD. III. *Diss. Acad. Ostrea*
Holsatica exhibens. 4to. 1708. with copperplates.

HARDERUS JOAN. JAC. VII. *Examen anat: Coch-*
leæ terrestris domiportæ. Basil. 1679. 8vo.
p. 73. with one plate.

Ant. Fel. Marsigli de Ovis Cochlearum,

cum Joh. Jac. Hardeeri Epistolis aliquot de Partibus. Genitalibus Cochlearum. Aug. Vindel. 1684. 8vo.

HATCHETT CHARLES. VIII. *Experiments on Shell and Bone,* in " Phil. Trans." 1799.

HEBENSTREIT Jo. ERN. IV. IX. *Museum Richterianum.* Lips. 1713. fol.

Dissertatio de Ordinibus Conchyliorum methodica Ratione instituendis. Lips. 1728. 4to. p. 28.

HERISSANT FRANCOIS DAVID. VIII. *Eclaircissemens sur l'Organisation jusqu'ici inconnue d'une Quantité considérable de Productions Animales, principalement de Coquilles des Animaux,* in the " Mém. de l'Acad. Roy. des Sc." 1766. p. 508—540. with plates.

HILL SIR JOHN. I. *History of Animals.* Lond. 1752. fol. with plates.

HOFER JOANNES. VII. *Observatio Zoologica,* in " Act. Helvet." vol. iv. p. 212, 213, tab. 9. fig. 21, 22. (1760.)

HOFFMAN Jo. FRIDERIC. II. V. *Dissertatiuncula de Cornu Ammonis nativo Littoris Bergensis in Norvegia,* in " Act. Acad. Mogunt." tom. i. p. 110—117. (1757.)

De Tubulis vermicularibus Cornua Am-

monis referentibus, in ".Act. Acad. Mogunt."
tom. ii. p. 16—20. (1761.)

*De Concha Sphærica fluviatili, alata, ex
badio et nigro Colore variegata,* in " Act.
Acad. Mogunt." tom. ii. p. 1—15. (1761.)

HUMPHREYS GEORGE. .VII. . *Account of the Gizzard
of the Shell called by Linnæus* ' BULLA LIG-
NARIA,' in " Trans. Linn. Soc." vol. ii.
p. 15—18. pl. 2. (1794.)

I

IMPERATO FRANCESCO. I. *Dell' Historia Naturale
di Ferrante Imperato Napolitano, Lib.* 28.
Neap. 1599. fol. with wooden cuts.

J

JACOBÆUS OLIGER. IV. *Museum Regium.* Haffn.
1696. fol. with copperplates.

JONSTON JOANNES. I. VI. *De exsanguibus Aquaticis,*
lib. iv. Amst. 1657. fol. with 20 copperplates.

Thaumatographia Naturalis, Amstel. 1632
et 1665. 16mo.

*A History of the Wonderful Things of
Nature.* London, 1657. fol.

K

KAMEL GEORGE JOSEPH. III. *De Conchis Turbini-*

bus Bivalvibus et Univalvibus Philippensibus,
in the "Phil. Trans." vol. xxv. p. 2396.
(1707.)

KAMMERER C. L. IV. *Die Conchylien in Cabinette der
Herrn. Erb-Prinzen von Schwartzburg-Ru-
dolstadt.* Rudolstadt, 1786. 8vo. with 12
plates.

*Nachtrag zu der Conchylien im Furstlichen
Cabinette zu Rudolstadt.* Lips. 1791. 8vo.
with plates.

KLEIN JA. THEOD. II. VIII. IX. XI. *Von Schaal-
thieren, Conchæ anatiferæ, Entenmuscheln,* in
" Abhandl. der Naturf. Gesellsch. in Dant-
zig." 2 theil. p. 349—354.

Tentamen Methodi Ostracologicæ. Lugd.
Batav. 1753. 4to. with 12 copperplates.

*Lucubratiuncula de Formatione, Cremento,
et Coloribus Testarum.*

Descriptiones Tubulorum marinorum. Gedan. 1731. 4to. with copperplates.

Commentariolum in Locum Plinii Hist. Nat.
(lib. ix. c. 33.) *de Concharum Differentiis in
Tent. Meth. Ostracologicæ.*

KNORR GEORG. WOLFFGANG. XII. *Les Délices des
Yeux et de l'Esprit, ou Collection générale
des différentes Espèces de Coquillage que la*

Mer renferme. Nuremb. tom. i. 1760.
tom. ii. 1773. 4to. with coloured plates.

Deliciæ Naturæ selectæ. Nuremb. 1766.
fol. vol. i. with 7 coloured plates.

KOELREUTER JOS. THEOPHILUS. II. VII. *Dentalii
Americani ingentis magnitudinis Descriptio,*
in " Nov. Comment. Ac. Petrop." 1766.
tom. xii. p. 352—356.

*Observationes Anatomico-Physiologicæ My-
tili Cygnei Linn. Ovaria concernentes,* in
" Nov. Sc. Imp. Petrop." tom. vi. p. 236—
239. (1790.)

KUNDMANN JOH. CHRISTIANUS. VI. *De Conchis et
Cochleis monstrosis pretiosisque,* in " Act. Ac.
Nat. Cur." vol. iii. p. 317. (1733.)

L

LAMARCK JEAN BAPT. IX. XII. *Observations sur les
Coquilles,* in the " Journ. d'Hist. Nat."
tom. ii. p. 269—280. (1792.)

*Prodrome d'une nouvelle Classification des
Coquilles,* in the " Mém. de la Soc. d'Hist."
Nat. de Paris," (An. 7.) p. 63—91.

Vers Testacés, in the " Tableau Encyclo-
pédique et Méthodique." (Paris, 1797-1798.)
4to. Livraison 62—64. 390 plates.

Système des Animaux sans Vertèbres.
Paris. (1801.) 8vo.

Histoire Naturelle des Animaux sans Vertèbres. Paris. tom. i. (1815.) 8vo.

LANGIUS CAR. NIC. IX. *Methodus nova Testacea marina in suas Classes, Genera et Species distribuendi.* Lucern. 1722. 4to. p..102.

LASKEY J. III. *Account of North British Testacea,* in " Memoirs of the Wernerian Natural History Society, vol. i. p. 370.

LAUERBNTZEN JOANNES. IV. *Auctarium Museii Regii.* Haffn. 1699. fol. with copperplates.

LEACH W. E. I. *Zoological Magazine.* Lond. (1815.) 8vo..with plates. .

LEDERMULLER MARTIN FROBENE. V. *Amusement Microscopique,* tom. i. Nuremb. 1764. 4to. pl. 4. 8. tom. ii. 1766. pl. 74.

LEEUWENHOEK ANTONIUS A. VIII. *De Ovis et Ovariis Testaceorum,* in " Phil. Trans." vol. xvii. (1694.) p. 593, 594; vol. xix. (1698.) p. 790—793; vol. xxvii. (1712.) p. 529—534.

LEGATI LORENZO. IV. *Museo Cospiano.* p. 92. Bologna, 1671. fol. with wooden cuts.

LE GENTIL. II. *Observations sur une Espèce de Varesch, &c.. et sur une petite. Coquille qui se*

loge dans le Tronc de cette Plante, in the
" Mém. de l'Acad. des Sc." 1788. p. 439—
442. tab. 20.

LEIGH CHARLES. III. *Natural History of Lanca-
shire, Cheshire, and Derbyshire.* Oxford,
1700. fol.

LESSERS FRED. CHRIST. I. *Testaceo-Theologia.* Lips.
1748 et 1756. 8vo. German, with copperplates.

LIGHTFOOT JOHN. III. *An Account of some minute
British Shells, either not observed or totally
unnoticed by Authors,* in " Phil. Trans."
vol. lxxvi. p. 160—170. (1786.) with 3 plates.

LINNÆUS CAROLUS. I. II. III. IV. IX. In *Mantissa
Plantarum altera.* Holm. 1771. 8vo.

Anomia descripta, in " Nov. Act. Soc.
Upsal." vol. i. p. 39—43. tab. 5. fig. 3.
(1773.)

Anomia patellæformis, in " Nov. Act.
Reg. Soc. Sc. Ups." vol. i. p. 42. fig. 6. 7.

Fauna Suecica. Lugd. Bat. 1746. 8vo.
Holm. 1761. 8vo.

Wästgöta Resa. Stockholm, 1747. 8vo.

Museum Tessinianum, Latine et Suecice.
Holm. 1753. with plates.

Museum Adolphi Frid. Suec. Regis, Lat. et Suec. Holm. 1754. fol. with plates.

Museum Ludovicæ Ulricæ Suecīæ Reginæ. Holm. 1764. 8vo.

Systema Naturæ. Ed. 1ma. Lugd. Bat. 1735. fol. Ed. 13ma. à Jo. FRID. GMELIN. (Lips. 1788.) 8vo. tom. i. pars 6.

System of Nature, by WILLIAM TURTON, M.D. part 1. 1802. 8vo.

Fundamenta Testaceologiæ. Resp. ADOLH. MURRAY. Upsal. 1771. 4to. p. 43. with 2 copperplates, and in *Amœn. Acad.* vol. viii. p. 107—150.

LINOCIER GEOFFROY. I. *Histoire des Poissons.* Paris, 1584. 12mo. with plates. Paris, 1619. 12mo. printed with his " Histoire des Plantès."

LISTER MARTIN. I. II. III. VII. *Historia, sive Synopsis methodica, Conchyliorum.* Lond. 1685-1692. fol. Oxon. 1770. à GULIELMO HUDDESFORD. with copperplates.

Observations concerning the odd Turn of some Shell Snails, &c. in " Phil. Trans." vol. iv. p. 1011. (1669.)

Historia Animalium Angliæ. Lond. 1678. 4to. with copperplates.

Appendix Hist. Anim. Angliæ. Ebor. 1681. 4to. with copperplates. Lond. 1685. 8vo.

Exercitatio Anatomica, in qua de Cochleis, maximè terrestribus, agitur. London, 1694. 8vo. with 6 copperplates.

Exercit. Anat. altera, in qua agitur de Buccinis fluviatilibus et marinis. Lond. 1695. with 6 copperplates.

Exercitatio Anatomica tertia. Lond. 1696. with 9 copperplates.

Anatomy of the Scallop, in " Phil. Trans." vol. xix. p. 567. (1697.) with 1 copperplate. Latin.

LOCHNER JOAN. HENR.   IV.   *Rariora Musei Besleriani.* 1716. fol. with copperplates.

LONICERUS ADAM.   I.   *Historiæ Naturalis Opus novum.* Francof. tom. i. 1551. fol. with wooden cuts, and tom. ii. 1555.

M

MACBRIDE DAVID.   VIII.   *On the Reviviscence of some Snails kept 15 Years,* in " Phil. Trans." vol. lxiv. p. 432—437. (1774.)

MAGNUS ALBERTUS.   I.   *De Animalibus.* Venet. 1495. fol.

MAJOR DANIEL. X. *Dictionarium Ostracologicum.*

MARSIGLI ANTONIO FELICE. VIII. *Relazione del Ritrovamento dell' Uova di Chiocciole.* Bologña, 1683. et Roma, 1695. 12mo. with 1 copperplate.

MARTIN THOMAS. XII. *Universal Conchologist.* London, vol. i. 1784; vol. ii. 1786. fol. with 160 plates.

MARTINI F. H. W. I. IX. *Neues Systematisches Conchylien Cabinet,* from vol. i. 1769 to vol. iii. 1777. 4to. continued by CHEMNITZ, with coloured plates.

MASSARIUS FRANCISCUS. XI. *Castigationes et Annotationes in* 9 *Librum Plinii de Nat. Hist.* Basil. 1537. 4to.

MATON WILLIAM GEORGE. II. III. In " Trans. Linn. Soc." vol. iii. p. 44-45. tab. 13. fig. 37-38. (1797.)

Observations relative chiefly to the Natural History, &c. of the Western Counties of England. Salisbury, 1797. 8vo. 2 vols.

Historical Account of Testaceological Writers. " Trans. Linn. Soc." vol. vii. p. 119.

MATTHIOLUS PETRUS. XI. *Commentaria in vi Dioscoridis.* Venet. 1554. fol. Latin, with plates.

MERRETT CHRISTOPH. III. *Pinax Rerum Natura-lium Britannicarum.* London, 1667. 8vo.

MERY JEAN. VII. *Remarques faites sur. la Moule des Etangs,* in " Mém. de l'Åcad. Franç." 1710. p. 408—426.

MESAIZE. II. *Observations sur les Conques anatifères,* in the " Magazin Encyclop." An. 2. tom. vi. p. 158. (1797.)

MOLINA GIOV. IGNAZIO. III. *Saggia sulla Storia Naturale del Chili.* Bologna, 1782. 8vo.

MOLL J. P. CARL. A. V. *Testacea Microscopica aliaque minuta ex Generibus Argonauta et Nautilus ad Naturam picta et descripta.* Wien. 1798. 4to. with 24 coloured plates. (Latin and German.)

MONTAGU S. III. *Testacea Britannica.* London, (1803.) 4to. with 16 coloured plates.

Supplement to the Same, (1808.) with 14 coloured plates.

MONTFORT DENYS DE. IX. *Conchyliologie Syste-matique.* Paris, (1808.) 8vo. 2 vols. with coloured plates.

MORAY SIR ROBERT. II. *Relation concerning Bar-nacles,* in " Phil. Trans." vol. xii. p. 925—927. with a figure. (1678.)

MORNICHEN ERICUS A. II. *Conchæ anatiferæ vin-*

dictæ. Resp. CLAUDIO URSIN. Haffn. 1697. 4to. with two plates.

MORTON JOHN. III. *Natural History of Northamptonshire.* London,.1712. fol. with plates,

MOSCARDO CONTE LODOVICO. IV. *Note del Museo del Conte L. M.* Pad. 1656. fol. with copperplates. Veron. 1672. fol. with plates and wooden cuts.

MULLER OTHO FREDERIC. III. VIII. IX. XII. *Zoologiæ Danicæ Prodromus.* Havniæ, 1776. 8vo.

Zoologia Danica. Havn. et Lips. vol. i. 1779. vol. ii. 1784. 8vo. Havn. 1781. fol. with copperplates.

On the same subject, in the " Journ. de Physique," (1777.) tom. xii. p. 111.

Vermium Terrestrium et Fluviatilium Historia. Havniæ, tom. i. 1773. tom. ii. 1774. 4to.

Animalium Daniæ et Norvegiæ rariorum ac minus notorum Icones. Havniæ, 1777. fol. with 40 plates.

MURRAY J. AND. VIII. *De Redintegratione Partium Cochleis Limacibusque Præcisarum.* Gotting. 1776. 4to. p. 19. and in his " Opuscula," vol. i. p. 315—342.

NEEDHAM TURBERVILLE. V. *An Account of some new Microscopical Discoveries.* London, 1745. 8vo. plate 6.

, NIEREMBERGIUS JOAN. EUSEBIUS. I. *Historia Naturæ.* Antverp. 1635. fol. with wooden cuts.

NORMANN LAURENCE. II. *Dissert. Acad. de Purpura.* Resp. EL. BASK. Upsal. 1686. 8vo. with wooden cuts.

O

OLEARIUS ADAM. IV. *Kunst-Gammer.* Slesv. 1666 and 1674. 4to. with plates.

OLIVI GIUSEPPE. III. *Zoologia Adriatica.* Bassano, 1792. 4to. with plates.

OLIVUS JOANNES BAPTISTA. IV. , *De reconditis et præcipuis Collectaneis in Mus. Calceolario asservatis.* Venet. 1584. 4to.

P

PALLAS P. S. I. *Miscellanea Zoologica.* Hag. Com. 1766. 4to. with plates.

Spicilegia Zoologica. Berol. 1767—1780. 4to. with copperplates.

Marina varia, nova, et rariora, in " Nov.

Act. Acad. Petrop." tom. ii. p. 229. tab. 7. (1787.)

PARSONS JAMES. II. *Observations on certain Shell-fish lodged in a large Stone brought from Mahon Harbour,* in " Phil. Trans." vol. xxv. p. 44—48. (1750.) with figures.

    Account of Pholas Conoides, in " Phil. Trans." vol. lv. p. 1—6. (1760.) with figures.

PENNANT THOMAS. II. III. *Anomia,* in " Nov. Act. Soc. Upsal.". vol. i. p. 38-39. fig. 4. (1773.)

    British Zoology, vol. iv. London, 1777. 8vo. with plates.

PETIVER JAMES. III. XII. *Account of Animals and Shells sent from Carolina,* in " Phil. Trans." vol. xxiv. p. 1951. (1705.)

    Pterigraphia Americana, in ejusdem " Gazophylacio."

    Description of some Shells found on the Molucca Islands, in " Phil. Trans." vol. xxii. p. 923—933. (1701.)

    Gazophylacium Naturæ et Artis. Lond. 1702—1711. fol.

    Opera Hist. Nat. spectantia. London, 1764. fol. 2 vols.

PLANCUS JANUS. III. *De Conchis minus natis.* Venet. 1739. et Rom. 1760. 4to. with plates.

De quibusdam Conchis minus notis, in
" Mem. sopra la Fisica," tom. i. (Lucca,
1743.)

PLINIUS SECUNDUS CAIUS. I. (Lived A.D. 80.)
Historia Mundi, Lib. ix.

PLOT ROBERT. III. *Natural History of Stafford-
shire*. Oxford, 1686. fol.

Natural History of Oxfordshire, 2d edit.
Oxford, 1705. fol. with figures.

POUPART FRANÇOIS. VIII. *Sur la Progression du
Limaçon aquatique dont la Coquille est
tournée en spirale conique*, in the " Journ.
des Sçavans," 22 Mars, 1694.

*Remarques sur les Coquillages à deux Co-
quilles, et premièrement sur les Moules*, in
the " Mem. de l'Acad. Roy. des Sc." 1706.
p. 52—61. with figures.

POWER HENRY. V. *Experimental Philosophy, Obs.* 1.
(London, 1664. 4to.)

PULTENEY R. I. *General View of Linnæus's Writings*,
edited by Maton. London, 1805. 4to. with
plates.

*Catalogues of the Birds, Shells, &c. of
Dorsetshire*, edited by Rackett. London,
1813.

R

REGENFUS FRANÇOIS MICHAEL. XII. *Choix de Co-quillages et de Crustacés.* Copenh. 1758. fol.

RETZIUS M. AND. II. IX. *Venus Lithophaga descripta,* in the " Mém. de l'Acad. de Turin," vol. iii. Corresp. p. 11—14. with fig.

Nova Testaceorum Genera. Resp. LAUR. MUNTER PHILIPSSON. Lund. 1788. 4to. p. 23.

RONDELETIUS GULIELMUS. I. *Universa Aquatilium Historia.* Lugd. 1555. fol. with wooden cuts.

ROUSSET. II. *Observations sur les Vers de Mer qui percent les Vaisseaux,* 2d edit. Haye, 1733. 8vo. with three plates.

RUMPHIUS GEORG. EVERARD. II. III. IV. XII. *De Nautilo remigante et velificante,* in " Eph. Ac. Nat. Cur." Dec. 2. An. 7. p. 8. (1689.) with fig.

De Ovo Marino; Porcellanis, seu Conchis Venereis, in " Eph. Ac. Nat. Cur." Dec. 2. An. 5. p. 222. (1686.) with plates.

D'Amboinsche Rariteitkamer. Amstel. 1705 et 1741. fol. with 60 plates.

Thesaurus Imaginum Piscium Testaceorum, quibus accedunt Conchylia, denique

Mineralia. Lugd. Bat. 1705, et Hag. Comitum, 1739. fol. with 60 plates.

RUTTY JOHN. III. *Essay towards a Natural History of the County of Dublin.* Dublin, 1772. 2 vols. 8vo.

RUYSCH FRIDERICUS. IV. *Thesaurus Animalium Primus.* (Latine et Belgice.) Amst. 1710. 4to. with plates.

S

SCHELHAMMER GUNTHER CHRISTOPH. II. VIII. *Conchæ Cochleæque recenter observatæ*, in " Eph. Ac. Nat. Cur." Dec. 2. An. 6. p. 212—216. (1688.) with figures.

Alia Cochlearum Genera itidem Mediterranea, Aquæ dulcis Incolis accensenda. Ibid. p. 216.

Animal in Cochlea minuta depressa degens, in " Eph. Ac. Nat. Cur." Dec. 2. An. 9. p. 245, 246. (1691.)

SCHLOTTERBECCIUS P. J. VIII. *Observationes de Cochlea quadam ad Turbines referenda*, in " Act. Helvet." vol. iv. p. 46—49. tab. 5. fig. 4. (1760.)

Observatio Physica de Cochleis quibusdam, &c. in " Act. Helv." vol. v. p. 275—288. tab. 3 A. 3 B. (1762.)

SCHONVELDE STEPH. VON. III. *Ichthyologia.* Hamb. 1624. 4to.

SCHRÖTER J. S. I. III. *Einleitung in die Conchylienkenntniss nach Linné,* 3 vols. 8vo. 1786. with plates.

Versuch einer systematischen Abhandlung über die Erdkonchylien um Thangelstadt. Berlin, 1771. 4to. with 2 plates.

Die Geschichte der Flussconchylien mit verzüglicher Rücksicht auf diejenigen welche in den Thüringischen wassern leben. Halle, 1779, with 11 plates.

Ueber den innern Bau des See und einiger ausländischen Erd - und Flust - schnecken. Frank. 1783. 4to. with 6 plates.

SCOPOLI JOH. ANT. I. III. *Introductio ad Historiam Naturalem.* Pragæ, 1777. 8vo. p. 386 —400.

Deliciæ Faunæ et Floræ Insubricæ. Ticini, 1786. with plates.

SEBA ALBERTUS. IV. *Descriptio Thesauri Rerum Naturalium,* tom. 3tius. (Amst. 1758. fol.) with plates.

SEGERUS GEORGIUS. IV. *Synopsis Methodica rariorum tum Naturalium tum Artificialium*

quæ in Museo D. Olai Wormii asservantur.
Hafn. 1653. 4to.

Sellius Godofred. II. *Historia Naturalis Teredinis seu Xylophagi marini.* Trajecta ad Rhen. 1733. 4to. with 2 plates. p. 353.

Shaw George. II. *Naturalist's Miscellany.* London, 1790. 8vo. with coloured plates.

Sibbald Sir Robert. II. III. IV. *Description of the Pediculus Ceti,* in " Phil. Trans." vol. xxv. p. 2314—2317. with a figure.

Scotia illustrata. Edinb. 1684. fol. with plates.

Letter to Dr. M. Lister, in " Phil. Trans." vol. xix. p. 321.

Auctarium Musei Balfouriani. Edinb. 1697. 8vo.

Slabber Martin. II. *Naturkundigne Verlustigingen.* (Haarl. 1778.) 4to.

Sloane Sir Hans. III. *Voyage to the Islands Madeira, Barbadoes, Nevis, St. Christopher's, and Jamaica,* vol. ii. p. 240-241. London, 1725. fol.

Smith Charles. III. *Antient and present State of the County and. City of Waterford.* Dublin, 1745, 8vo. 1774, 8vo.; with many additions.

Antient and present State of the County of Cork. Dublin, 1750, 8vo. 2 vols.

Antient and present State of the County of Kerry. Dublin, 1756, 8vo.

SOLDANI A. I. V. *Testaceographia et Zoophyto-graphia parva ac microscopica.* Senis, 1789. 3 parts. fol. with 179 plates.

SPALLANZANI LAZARO. VIII. *Risultati di Esperienze sopra la Reproduzione della Testa nelle Lumache Terrestri,* in " Mem. della Soc. Ital." (1782.) tom. i. p. 581—612. (1784.) tom. ii. p. 506—602. with fig.

STENO NICOLAUS. VIII. *De Solido intra Solidum naturaliter contento Dissertationis Prodromus.* Flor. 1669. 4to.

T

TOURNEFORT JOSEPHUS PITTON. IX. *Introductio ad Historiam Testaceorum,* in " Ind. Test." NIC. GUALTIERI.

TURTON W. III. IX. *General System of Nature, by Sir Charles Linné, translated from Gmelin, &c.* London, 1806. 7 vols. 8vo.

British Fauna. vol. i. Swansea, 1807. 12mo.

V

VALENTINI MICH. BEEN. IV. VII. *Museum Museorum* (Germanice). Francf. ad Mœn. 1704. fol. 2 vols. with plates.

Amphitheatrum Zootomicum. Francof. 1720. fol. with plates.

VALENTYN FRANCOIS. III. *Oud en Nieuw Oost-Indien.* Dordrecht et Amst. 1724—1727. tom. viii. fol. with plates.

Verhandeling der Zee-horenkens en Zee-gewassen in en omtrent Amboina on de nabygelegne Eilanden. Amst. 1754. fol. with the same plates.

VALLISNERI ANTONIO. II. *Insetti marini analoghi alle Patelle o cimici degli Agrumi*, in " Opere Fisico-mediche," tom. ii. p. 95.

VAN DER WIEL CORNELIUS STALPART. II. In his *Observationes rariorum Med. Anat. et Chirurg.* Cent. post. p. 458—469. with plates. (Lugd. Bat. 1727. 8vo.)

VAUQUELIN. VIII. *Observations Chimiques et Physiques sur la Respiration des Insectes et des Vers*, in the " Ann. de Chim." tom. xii. p. 273—291. (1792.)

VICENTIUS. I. *Speculum Naturale.* Venet. 1495. folio.

VINCENT LEVEN. IV. *Wondertoonel der Natur.* Amst. 1706. 4to. with plates.

 Description abrégée, &c. Amst. 1715. 4to. with the same plates.

W

WALKER GEORGE. III. V. *Testacea minuta rariora nuperrimè detecta in Arena Littoris Sandvicensis.* Londini, 1784. 4to. with three plates.

WALLACE. III. *Account of the Islands of Orkney,* 2d edit. Lond. 1700. 8vo.

WALLIS JOHN. III. *Natural History and Antiquities of Northumberland,* vol. i. p. 366. (Lond. 1769. fol.)

WHYTT ROBERT. VII. *Description of the Matrix, or Ovary, of the B. ampullatum,* in " Edinb. Phys. and Lit. Essays," vol. ii. p. 8. (1756.) with fig.

WILLIS THOMAS. VIII. *Exercitationes duæ de Anima Brutorum.* (Lond. 1672.) 8vo. tab. 2da.

WITSEN NICHOLAS. III. *Description of certain Shells found in the East Indies,* in " Phil. Trans." vol. xvii. p. 870. (1693.) with figures.

Wood William. IX. *Observations on the Hinges of British Bivalve Shells*, in " Trans. Linn. Soc." vol. vi. p. 154. t. 14. 15. 16. 17. 18. (1802.)

General Conchology, vol. i. Lond. 1815. 8vo. with coloured plates.

Zoography. (1807.) 8vo. vol. ii. with plates.

Index Testaceologicus. London, 1818. 8vo.

Wormius Olaus. IV. *Museum Wormianum*, cap. vi. vii. viii. Lugd. Bat. (1655.) fol. with one figure.

THE " Nomenclature," which is inserted immediately after the " Introduction," refers to the *Generic* and *Specific* Characters of Shells, among which Colours ought not, properly speaking, ever to be ranked: they distinguish *varieties* and *individuals* which differ in the same species, and, being liable to uncertainty and change more than any other feature, are to be considered as the last in value. To the last place, therefore, they have been consigned in the present work, which has been arranged as much as possible upon the regular gradation necessary to the description of any subject of Natural History, according to the spirit of the Linnæan System.

In the following list those colours only are enumerated, which occur in the Latin descriptions of Testacea. The name of a well known pigment, or some other permanent standard, is affixed to those respecting which there can be any doubt. Wherever the termination *ish* is used, it implies that an inconsiderable portion of the colour tinges white, or enters into combination with the other mentioned;—as redd*ish* yellow, signifies that yellow is the principal, and red but a small ingredient in the hue.

TABLE OF COLOURS.

ALBUS	WHITE LEAD.	
Candidus	*Pure White*	
Albidus	*Whitish*	
Niveus	*Brilliant White*	Snow.
Lacteus	*Dilute White*	Milk.
Cærulescente-albus	*Bluish White*	
Flavescente-albus	*Yellowish White*	
Virescente-albus	*Greenish White*	
Purpurascente-albus	*Purplish White*	
Rufescente-albus	*Reddish White*	
Fuscescente-albus	*Brownish White*	

ATER	BLACK	Lamp Black.
Niger	*Black*	Indian Ink.
Nigricans	*Blackish*	

CÆRULEUS	BLUE	Indigo.
Cærulescens	*Bluish*	
Virescente-cæruleus	*Greenish Blue*	
Chalybeus	*Purplish Blue*	Steel spring.
Cyaneus	*Light Blue*	Azure, Sky.

FUSCUS	BROWN	Bistre.
Fuscescens	*Brownish*	
Cærulescente-fuscus	*Bluish Brown*	

Atro fuscus	*Black Brown*	Cologn Earth.
Pullus	*Russet Brown*	Bark of the Oak.
Brunneus	*Chesnut Brown*	Span. Chesnut.
Fulvus	*Fulvous*	Tawny.
Cinnamoneus	*Yellowish Brown*	Cinnamon.
Spadiceus	*Chesnut*	Terra Sienna.
Badius	*Bay*	Horse Chesnut.
Ferrugineus	*Ferruginous*	Rust.
Obscŭrus	*Dusky*	Sepia.
Hepaticus	*Purplish Brown*	Liver.
Mustelinus	*Pale Tawny*	Weasel.
Coffeus	*Dark Brown*	Coffee.

GRISEA	GRAY	
Cinereus	*Cinereous*	Ashes.
Cinerascens	*Grayish*	
Rubello-cinereus	*Reddish Gray*	
Virescente-canus	*Greenish Gray*	
Cærulescente-ci- nereus	*Bluish Ash*	
Plumbeus	*Plumbeous*	Lead.
Murinus	*Black Gray*	Mouse.

LUTEUS	YELLOW	Gamboge.
Flavus	*Bright Yellow*	Patent Yellow
Flavescens	*Yellowish*	
Viridescente-flavus	*Greenish Yellow*	
Stramineus	*Pale Yellow*	Straw.
Ochroleucus	*Ochraceous*	Yellow Ochre.
Aurantius	*Red Yellow*	Orange.
Croceus	*Reddish Yellow*	Saffron.
Corneus	*Brownish Yellow*	Horn.
Cereus	*Dull reddish Yellow*	Bees'-wax.
Aureus	*Brilliant Yellow*	Gold.
Citrinus	*Pale Yellow*	Citron.

R

PURPUREUS	**PURPLE**	
Purpurascens	*Purplish*	
Rufescente-purpur.	*Reddish Purple*	
Violaceus	*Bluish Purple*	Violet.
Janthinus	*Deep Violet*	
RUFUS	**RUFOUS**	Deep Red.
Ruber	*Red*	Vermilion.
Rubellus	*Reddish*	
Sanguineus	{ *Sanguineous* *Crimson*	Lake.
Rubicundus	*Pale Red*	
Rutilus	*Bright Red*	Red Lead.
Vinosus	*Purplish Red*	Claret.
Testaceus	*Dull brownish Red*	Tile.
Flammeus	*Bright Yellow Red*	Flame.
Roseus	*Rosy*	Dilute.
Coccineus	*Scarlet*	
VIRIDIS	**GREEN**	Sap Green.
Virescens	*Greenish*	
Flavicante-viridis	*Yellowish Green*	
Porraceus Prasinus	} *Yellowish Green*	Leek.
Olivaceus	*Brownish Green*	Olive.
Thalassinus	*Bluish Green*	Sea.
Glaucus	{ *Glaucous* *Whitish Green*	

LIST OF PLATES.

THE END.

LONDON:
PRINTED BY J. MOYES, BOUVERIE STREET.

This Day is Published,

The MODERN TRAVELLER; or a Popular Description, Geographical, Historical, and Topographical, of the various Countries of the Globe, compiled from the latest and best Authorities (to be continued in Monthly Parts), embellished with correct Maps and numerous Engravings, price 2s. 6d. each.

Already Published,

PALESTINE, or the HOLY LAND, forming Parts I. and II., embellished with a Map and Three Plates, and may be had in One Volume, very neatly half-bound and lettered, price 6s.; or calf gilt extra, price 7s.

SYRIA and ASIA MINOR, forming Parts III. to VI., with Two Maps and Six Plates, making Two Volumes, price as above.

BRAZIL and BUENOS AYRES, forming Parts VII. to X., with a Map and Seven Plates, making Two Volumes, price as above.

MEXICO and GUATIMALA, forming Parts XI. to XIV., with a Map and Six Plates, price as above.

COLOMBIA, forming Parts XV. and XVI., with a Map and Three Plates, price as above.

ARABIA, RUSSIA, SPAIN, PERU, CHILI, GREECE, EGYPT, &c. &c. will follow in succession.